T0271856

Bioinformatics for Oral Cancer

Amid the rising global concern of oral cancer, this book provides a compelling exploration of the intricate oral cavity, focused on shedding light on early diagnosis and addressing outdated paradigms, it delves into the persistent challenges of oral premalignant lesions. Tailored for both beginners and researchers, its six chapters encompass the spectrum of genome sequencing, diagnostic biomarkers, gene expression, and more. Discover a fusion of basic and clinical sciences, aiming to invigorate the study of bioinformatics and oral cancer, and ultimately improve survival rates.

Bioinformatics for Oral Cancer: Current Insights and Advances serves as a comprehensive guide, offering a deep dive into the multifaceted landscape of oral cancer research and bioinformatics. Within its pages, readers will uncover a wealth of knowledge, starting with foundational chapters introducing bioinformatics and establishing the backdrop of oral cancer. The book then progresses into the realm of diagnostic biomarkers, revealing cutting-edge methodologies for their identification in the context of oral cancer. The book's keen focus extends to gene expression profiles and the intricacies of gene sequencing in the context of oral cancer progression. By systematically unravelling these critical aspects, the book bridges the gap between basic and clinical sciences, equipping readers with a holistic understanding of bioinformatics' pivotal role in enhancing our grasp of oral cancer's complexities.

By deciphering the enigmatic landscape of oral premalignant lesions, the book equips clinicians and researchers with tools to predict malignant potentials. Its meticulous exploration of gene expression profiles and sequencing promises to reshape early detection strategies, propelling the field towards improved diagnosis and treatment outcomes.

Bioinformatics for Oral Cancer
Current Insights and Advances

Edited by
Mahesh KP, Raghavendra Amachawadi,
Shiva Prasad Kollur and Chandan Shivamallu

CRC Press
Taylor & Francis Group
Boca Raton London New York

CRC Press is an imprint of the
Taylor & Francis Group, an **informa** business

Designed cover image: Shutterstock, trailak amtim

First edition published 2024
by CRC Press
2385 Executive Center Drive, Suite 320, Boca Raton, FL 33431

and by CRC Press
4 Park Square, Milton Park, Abingdon, Oxon, OX14 4RN

CRC Press is an imprint of Taylor & Francis Group, LLC

ISBN: 9781032625706 (hbk)
ISBN: 9781032625669 (pbk)
ISBN: 9781032625713 (ebk)

DOI: 10.1201/9781032625713

Typeset in Minion Pro
by KnowledgeWorks Global Ltd.

Contents

Preface

THE CONTENT OF THIS book is so accurately portrayed by every word of its title. It does indeed confine itself to the oral cavity, so do not expect to see oropharyngeal cancers, suddenly so prevalent. Oral cancer in humans has emerged as significant public health challenge globally, particularly in countries of Southeast Asia. Although the oral cavity is easily accessible as is the population at risk, early diagnosis has been painfully slow when compared to the enhanced early detection of cancers of the breast, colon, prostate, and melanoma. As a result, the mortality rate from oral cancer for the past four decades has remained high, at over 55%, in spite of advances in treatment modalities. This contrasts with a considerable decrease in mortality rates for cancers of the breast, colon, prostate, and melanoma during the same period. In spite of increased diligence on the part of clinicians in their examination of patients at risk, early diagnosis of oral cancer continues to be impeded and elusive because of the persistence of outdated paradigms, and the lack of an easily available diagnostic adjunct. This is particularly evident in the persistent challenge of deciphering the malignant potentials of the various oral premalignant lesions (OPLs). To increase the early detection of oral cancer with the attendant increase in survival rates, and OPLs with the likelihood of transition to oral cancer, there is the need to identify diagnostic screening modalities that accurately predict the malignant potentials of OPLs. This book, titled *Bioinformatics for Oral Cancer: Current Insights and Advances* is intended for the introduction to bioinformatics and better understanding of mouth health for researchers and four-year colleges. Further, it provides a comprehensive, yet reference-friendly, update encompassing the spectrum of tools used in genome sequencing, management philosophies, and progression indicators of oral cancer. For convenience of reference, the book has been divided into six chapters. Chapters 1 and 2 cover the introduction to bioinformatics and background of oral cancer, respectively. Chapters 3 and 4

emphasize biomarkers as diagnostic tools and the methodologies for identifying them in oral cancer scenario. Chapters 5 and 6 point out the gene
expression profile and gene sequencing in oral cancer. The book enhances
integration of basic and clinical sciences. Our goal of the book is to make
studies in bioinformatics and oral cancer stimulating and exciting.

Editor Biographies

Dr. Mahesh K P is an Associate Professor specializing in oral diagnostics and radiology at JSS Dental College and Hospital, JSS AHER, Mysuru, India. With a teaching journey spanning 17 years, dedicated to both educating and researching in diverse areas of dentistry. Achievements include receiving the Asia Pacific Young Achiever Award (Education Sector) from Belgave and the Commonwealth Vocational Services (Education Sector) recognition in 2019, authoring 45 publications, securing three patents, and authoring two textbooks along with a textbook chapter in the field of dentistry.

Raghavendra Amachawadi, also known as Raghu, holds the position of an Associate Professor of Food Animal Therapeutics, Department of Clinical Sciences at the College of Veterinary Medicine, Kansas State University. Raghu's journey began with a BVSc degree from the Veterinary School in Bangalore, India, which is comparable to a DVM degree. In 2008, a new chapter began by coming to the United States earning from Kansas State University an MS degree in 2010, followed by a PhD in Microbiology/Epidemiology in 2014.

Shiva Prasad Kollur is an Associate Professor in the field of Chemistry at Amrita Vishwa Vidyapeetham, located in Mysuru, India. Academic pursuits have centred around the realms of Bioinorganic and Materials Chemistry. Professor Kollur has contributed significantly to the scientific community, authoring and publishing over 120 research articles in prestigious international peer-reviewed journals as well as successfully patented research discoveries. Another gratifying aspect has been the role of mentor, having the privilege of guiding more than 40 scholars, both at the graduate and postgraduate levels, in their pursuit of academic accomplishments.

Chandan S, Deputy Dean of Research and Associate Professor at JSS AHER, is an accomplished researcher with an exceptional 15-year track record in the world of academia. His journey is marked by numerous accolades, including the highly prestigious Young Scientist Award conferred by the Department of Science and Technology. He has authored an impressive 248 international publications, reflecting his unwavering commitment to advancing knowledge in his field. Beyond publications, he has also secured five patents, underlining his innovative contributions to research. Importantly, his work has gone beyond theory, resulting in the creation of three prototypes. Mentorship and education are integral aspects of Chandan S's career. He has successfully guided four PhD students. Currently, he continues to mentor six more students, further highlighting his commitment to fostering academic excellence. Professor Chandan S's work exemplifies a multifaceted commitment to research, mentorship, and innovation. His diverse research interests and collaborative efforts not only enrich the academic community but also contribute significantly to the broader advancement of knowledge in his field. His impressive career, complemented by a plethora of awards and recognitions, solidifies his status as a leading figure in research and academia.

Contributors

Bhargav Shreevatsa
Department of Biotechnology and Bioinformatics
JSS Academy of Higher Education and Research
Mysuru, Karnataka, India

Kavana C P
Department of Biotechnology and Bioinformatics
JSS Academy of Higher Education and Research
Mysuru, Karnataka, India

Chandan Dharmashekara
Department of Biotechnology and Bioinformatics
JSS Academy of Higher Education and Research
Mysuru, Karnataka, India

Anisha S Jain
Department of Microbiology
JSS Academy of Higher Education and Research
Mysuru, Karnataka, India

Abhigna N
Department of Biotechnology and Bioinformatics
JSS Academy of Higher Education and Research
Mysuru, Karnataka, India

Sumitha E
Department of Biotechnology and Bioinformatics
JSS Academy of Higher Education and Research
Mysuru, Karnataka, India

Bhavana H H
Department of Microbiology
JSS Academy of Higher Education and Research
Mysuru, Karnataka, India

Mahesh K P
Department of Oral Medicine and Radiology
JSS Dental College and Hospital
JSS Academy of Higher Education and Research
Mysore, Karanataka, India

Shiva Prasad Kollur
School of Physical Sciences
Amrita Vishwa Vidyapeetham
Mysuru, Karnataka, India

Sai Chakith M R
Department of Pharmacology
JSS Medical College
JSS Academy of Higher Education and Research
Mysuru, Karnataka, India

Ashwini Prasad
Department of Microbiology
JSS Academy of Higher Education and Research
Mysuru, Karnataka, India

Viveka S
Department of Oral Medicine and Radiology
JSS Dental College and Hospital
JSS Academy of Higher Education and Research
Mysore, Karanataka, India

Siddesh V Siddalingegowda
Department of Microbiology
JSS Academy of Higher Education and Research
Mysuru, Karnataka, India

Chandan Shivamallu
Department of Biotechnology and Bioinformatics
JSS Academy of Higher Education and Research
Mysuru, Karnataka, India

Umamaheshwari S
Department of Microbiology
JSS Academy of Higher Education and Research
Mysuru, Karnataka, India
Karnataka

Author Contributions
The authors involved in the investigation, namely BS, CD, ASJ, VS, BHH, AN, AP, SVS, KCP, SCMR, CS, US, SPK, and MKP; authors involved in writing—original draft preparation—include CS, SPK, RA, and MKP. All authors have read and agreed to the published version of the manuscript.

Data Availability Statement
All the data originated from this book is available from the authors under request.

Acknowledgments
The authors CS, MKP, VS, BS, CD, ASJ, KCP, SCMR, BHH, SVS, US, AP and AN acknowledge the support and infrastructure provided by JSS AHER. SPK thank Amrita Vishwa Vidyapeetham for support extended towards the collaborations. RA acknowledge Kansas State University, KS, USA for their support.

Conflicts of Interest
The authors declare no conflict of interest.

CHAPTER 1

Introduction to Bioinformatics

Viveka S[1], Bhargav Shreevatsa[2+], Kavana CP[2],
Chandan Dharmashekara[2], Mahesh KP[1],
Shiva Prasad Kollur[3], Chandan Shivamallu[2]

[1]Department of Oral Medicine and Radiology,
JSS Dental College and Hospital, JSS Academy of Higher
Education & Research, Mysuru, Karnataka, India

[2]Department of Biotechnology and Bioinformatics, JSS Academy
of Higher Education and Research, Mysuru, Karnataka, India

[3]School of Physical Sciences, Amrita Vishwa Vidyapeetham,
Mysuru, Karnataka, India

INTRODUCTION

Bioinformatics is a field that has grown in importance in the field of bio-medical research. Its role includes deciphering genomic, transcriptomic, and proteomic data generated by high-throughput experimental methods, as well as organising data from conventional biology. Bioinformatics is a novel concept that focuses on solving biological problems using information science. Mostly it deals with data collection, storage, retrieval, and analysis from databases (Bayat 2002). In the last decade, it has catalysed research in the field of healthcare to a great extent. Bioinformatics can aid research in dentistry by understanding the underlying pathways and mechanisms in certain oral diseases. It also helps in the early prediction and personalised treatment of cancer. It can also assist in developing patient care databases. Image processing of X-rays and CT can supplement

DOI: 10.1201/9781032625713-1

the diagnosis (Singaraju, Prasad, and Singaraju 2012). For both the clinical practice of dentistry and the dental research community, oral illnesses and disorders represent prospective study areas. The development of tools (statistical and computational) that can help in the understanding of the mechanisms underlying biological difficulties in the research is still necessary for bioinformatics, nonetheless (Kuo 2003).

Oral malignancies are one of the most prevalent types of cancer worldwide. Oral squamous cell carcinomas (OSCCs) account for more than 90% of all cancers. Visual examination alone, on the other hand, may lead to the overlooking of minor lesions and the failure to distinguish between malignant and benign oral diseases. As a result, OSCC is detected at an advanced stage, lowering the chance of survival (Pires et al. 2013). Early detection of lesions at risk of malignant transformation, as well as the discovery of lesions at risk of malignant transformation, is critical for reducing the need for major surgery and enhancing disease-free survival. Saliva has grown in prominence as a convenient source of biomarkers for detecting oral and systemic diseases (Piyarathne et al. 2021). The tight connection between oral dysplastic/neoplastic cells and saliva makes saliva an attractive choice for non-invasive and very accurate diagnostic testing. Studies have focused on the examination of bodily fluids, sometimes known as "liquid biopsy," for the discovery of OSCC diagnostic and prognostic biomarkers in the previous decade. Because of its closeness to cancer cells, non-invasive collection, accessibility, and cost-effective sample, saliva has grabbed researchers' attention (Borse, Konwar, and Buragohain 2020).

The identification of several potential biomarkers that express metabolic or proteomic activity, as well as epigenetic and genomic alterations of the malignant cells, has benefited from advances in our understanding of OSCC growth at the molecular level and has favoured the early detection and diagnosis of a feasible pathological state. Recent findings indicate that the inclusion of "salivary liquid biopsy" has significant potential for OSCC diagnosis and administration, as it improves prognosis, treatment, and follow-up, even though standardised techniques for accurate biomarkers are still being developed. The majority of patients with head and neck cancer are still managed based on an evaluation of the macroscopic tumour features and severity of the disease (Yakob et al. 2014). However, it might be able to foretell which people require severe treatment and which do not if genomic data could be utilised to guide treatment choices. With more accuracy, surgeons might be able to tell good tissue from malignant tissue, preventing deformity and functional loss

(Abdurakhmonov 2016). The identification of many illnesses and infections that cause disorders in the head and neck region is made possible by comparative proteomics. There have already been a few proteomic studies in the field of oral pathology that have helped to identify molecular risk factors and therapeutic targets (Berger and Mardis 2018).

Applications of Bioinformatics

We will gain a better understanding of mouth health, craniofacial growth and deformity, and the pathophysiology of oral problems thanks to these methods of extensive genetic analysis. The notion is that by integrating dental clinical specialties and dental basic sciences, "informatics," including dental informatics and bioinformatics, may build a bridge of opportunity to generate new hypotheses and ideas. Dental informatics and bioinformatics will need to work together to apply genomic data, both clinical and genomic, as both fields face similar methodological difficulties in understanding how the enormous amounts of data are transformed into a better overall understanding and more opportunities for application to oral diseases. Furthermore, finding genes for novel biomarkers and therapeutic targets would be made simpler by improved integration of genomes, transcriptomes, proteomics, and proteomes as they respond to illness and various chemicals (Garcia, Kuska, and Somerman 2013).

Repetitive sequence detection, regulatory elements, restriction analysis, primer design, protein sequence extraction from nucleotide sequence to predict structure, estimation of the evolutionary distance for phylogeny reconstruction, determination of the active sites of enzymes, identification of mutations, characterisation of SNPs (haplotypes), and other topics are currently the focus of bioinformatics.

Our understanding of biological processes is improved by bioinformatics. Bioinformatics differs from other approaches in that it focuses on creating and using computationally demanding tools to accomplish this goal. Sequence alignment, gene discovery, genome assembly, drug discovery, drug design, gene expression, protein structure prediction, protein–protein interaction prediction, genome-wide association studies, and evolution modelling are some of the major research projects in this area (Figure 1.1) (Oulas et al. 2019).

Historical and Conceptual Vision of Bioinformatics

Ben Hesper and Paulien Hogeweg first used the word "bioinformatics" in 1970. Ten years before DNA sequencing became an affordable option,

FIGURE 1.1 Applications of bioinformatics in system biology.

bioinformatics first emerged. Historical turning points that could be highlighted for its development include the 1953 release of Watson and Crick's structure of DNA as well as the 1960s data collecting and understanding of biochemistry and protein structure with Pauling, Coren, and Ramachandran's work. Because of her efforts to organise information about protein three-dimensional (3-D) structure, Margaret O. Dayhoff is referred to as the "Mother of Bioinformatics." A vast number of protein sequences from various creatures needed to be manually analysed and compared, but this proved to be unfeasible, necessitating the development of new computational approaches. The first collection of "Protein Information Resources" was put together by Margaret Oakley Dayhoff and her colleagues at the National Biomedical Research Foundation

(PIR) (Gauthier et al. 2019). The protein sequences were successfully categorised by Dayhoff's team into discrete groups and subgroups using matrices based on sequence similarity and percent acceptable mutation. This was released as a protein sequence atlas, which is frequently used in database similarity searches and protein sequence alignments. This strategy laid the groundwork for molecular evolution and protein sequence alignment. This is because of its role in the development of tools like peptide sequence calculators, X-ray crystallography software, and computational approaches for protein sequence comparison that let us infer evolutionary links across different kingdoms. Elvin A. Kabat made additional contributions to the bioinformatics field in the 1970s with his extensive protein sequence analysis of large databases of antibody sequences (Pearson 2013).

To provide immunology investigations a theoretical foundation, George Bell and colleagues began compiling DNA sequences into GenBank in 1974. From 1982 to 1992, Walter Goad's team worked on the initial edition of GenBank, and as a result of their efforts, the DNA sequence databases known as GenBank, "The European Molecular Biology Laboratory, and DNA DataBank of Japan were developed in 1979, 1980, and 1984, respectively. The most important advancement in DNA sequence databases was the addition of web-based search algorithms that enable researchers to locate and contrast target DNA sequences (Buckner et al. 2016). The original "GENEINFO" and its derived version, "Entrez," were created by David Benson, David Lipman, and other collaborators. Researchers could swiftly scan database-indexed sequences using this software and compare them to a query sequence. The software was easily accessible thanks to the database's web-based interface from the National Center for Biotechnology Information (NCBI). The analysis, comparison, and visualisation of molecular sequences have become easier, and a variety of approaches have helped bioinformatics advance in this area. Examples of such developments include the creation of dot matrix and diagram methods, sequence alignment through dynamic programming, the discovery of local alignments between sequences, multiple sequence alignment tools, the prediction of RNA secondary structures, the identification of sequence evolutionary relationships, and the assignment of gene function based on sequence similarity to known functions from models. The development of FASTA, BLAST, and their numerous variants has advanced the science of bioinformatics and significantly enhanced biological data analysis. The field of bioinformatics has made major strides, including the establishment of web-based genome databases for many prokaryotic and eukaryotic organisms and tools for predicting potential protein sequences, structure, and function of proteins and genes based on DNA sequences (Manzoni et al. 2018).

The development of bioinformatics to this point would not have been possible without the advancement of computers, which has been made possible by the contributions of many other scholars. The enormous progress we have made today is mostly due to advances in computing power and genome programmes. Large-scale capillary DNA sequencers and fluorescent labelling of dideoxynucleosides made it possible to gather a tonne of data in the 1990s (Prosdocimi, 2010). The quantity of data and the number of entire genomes are growing as a result of the advancement of next-generation sequencing (NGS) technology (Gupta and Verma 2019).

As a result, computers must be used in study to comprehend genetic diversity as well as the evolutionary and functional principles that underpin genetic architecture. The very next stage could be to investigate entire genomes, transcriptomes, or metabolomes independently and instead computationally model full living creatures and their environments while concurrently accounting for all molecular categories (Xu and Xu 2004).

Tools and Techniques in Bioinformatics

One of the most widely used platforms for microarray research is Bioconductor, an open-source and open development software project built on the R programming language. Arrays can be used to classify and discover gene profiles linked to a variety of malignancies, including oral cancer. This kind of genetic approach will help researchers better understand disease development, which will lead to better diagnosis and therapy for patients. Based on their expression profile, it can help discover and classify genes linked to malignancies, periodontal disorders, and cavities. Bioinformatics analysis is a crucial step in the process of analysing the data generated from microarray analysis (Dalal and Atri 2014).

Biological Database

Online databases are used to store the massive volumes of data created. A biological database is a piece of computer software or a website that

TABLE 1.1   Commonly Used Domain-specific Database

Database	Function
GenBank	Nucleotide sequence
Protein Data Bank	Three-dimensional macromolecular structures
ArrayExpress Archive	Functional genomics data
UniProt	Sequence and functional information on proteins
EMBL (European Molecular Biology Laboratory)	DNA database
Gene Expression Omnibus (GEO)	Functional genomics data repository
National Cancer Institute	Cancer gene expression profiles
Human Protein Atlas (HPA)	Expression profiles of human protein coding gene
Swiss-Prot	Protein knowledgebase
Swiss-model	Protein structure models
Reactome	Human biological pathways
Kyoto Encyclopedia of Genes and Genomes	Biological pathways and metabolism

allows you to edit, query, and retrieve data. Primary and secondary databases are the two types of databases. Experimental data such as nucleotide sequences, protein sequences, and macromolecular structures are stored in primary databases. Example: Protein Databank. Results produced from main data are stored in secondary databases. Ensemble, for example, GenBank, SwissProt, and Protein Data Bank are some of the most regularly utilised databases. Each database is dedicated to a single function (Table 1.1) (Wang, Wu, and Cai 2018).

Alignment

By comparing the unknown sequence with one or more known sequences, the common portions are predicted. Natural selection tends to conserve identical residues with comparable functional and structural roles during evolution. The ideal alignment aligns two or more sequences so that as many identical or comparable residues as possible are matched. Nucleotide sequences, such as DNA or RNA, or amino acid sequences are examples of sequences (Chatzou et al. 2016).

The two types of sequence alignment are multiple sequence alignment (MSA) and pairwise sequence alignment (PSA). While MSA aligns more than two similar sequences at once, PSA only does it for two. MSA is better than PSA since it takes into account many members of a sequence family and so delivers more biological information. Proteins are important biological molecules that hold structural and functional information, hence amino acid sequence alignment is more important. On sequences of equal length, global alignment is conducted, and alignment must be completed over the length (Fiser 2010).

Protein Structure and Function Prediction

Bioinformatics approaches for protein structure prediction include sequence similarity searches, MSA, domain identification, secondary structure prediction, solvent accessibility prediction, fold recognition, 3D model construction, and model validation. The three fields of structure prediction are ab initio prediction, fold identification, and homology modelling. The ab initio approach uses physics and chemistry to predict the structure of a protein. The process of predicting the structure of proteins for which there is no equivalent structure in the Protein Database (PDB), but for which local sequence and structural relationships involving brief protein fragments, as well as secondary structure prediction, are included in the prediction process, is known as "ab initio" or "de novo" (Brem et al. 2007).

Using the three-dimensional structures of one or more similar proteins, a procedure known as homology modelling can be used to predict the three-dimensional structure of a protein. Fold recognition is the process of determining which of a library of known structures has a folding pattern that matches a query protein with a known sequence but unknown structure. Fold recognition is the process of finding structures in the PDB that are comparable to an unknown sequence but are difficult to identify. The structure of a protein determines its function. Detecting homology or employing structural templates obtained from enzyme active sites can be used to predict function from structure. It is also possible to look for sequence-conserved patches, clefts, and electrostatic potentials on the protein surface (Vodkin et al. 2004).

Data Analysis

In the biomedical research community, microarray technologies have become a standard tool for evaluating gene expression patterns on a worldwide scale. Instead of analyzing individual genes, researchers may look at the expression of thousands of genes at the same time. Microarray techniques were first used to examine differential gene expression in complex RNA populations. Microarrays are made up of unique DNA sequences called probes that are biochemically attached to a glass slide. The "target" is the mRNA from the tissue of interest that is labelled. Because "mRNA" degrades quickly, it must be transformed into a more stable complementary DNA form (cDNA). The "target" that will be placed onto the glass slide for hybridisation is the converted mRNA to cDNA. On DNA microarrays, hybridisation is a process in which complementary sequences bond to each other under the right circumstances. The base pairs "A" and "T" are complementary, while "C" and "G" are complementary. The robotically spotted cDNA arrays and the short photolithographic oligonucleotide array are the two most often used microarray systems, which differ in design, technique, and analysis. Long oligonucleotide arrays, a mix of the two platforms, have lately acquired traction: The probes range in length from 30 to 80 base pairs (Alzubaidi et al. 2021).

For cDNA microarrays, polymerase chain reaction (PCR) products must be robotically printed onto a glass slide in a two-dimensional grid. Expressed sequence tags (ESTs), which have lengths ranging from 100 to 1000 base pairs, are amplified into double-stranded sequences by PCR. There are more than 35,000 cDNAs that can be found on a glass slide. cDNA arrays employ a two-fluorescent-dye method, in which the two RNA samples are labelled with either a Cy5 or a Cy3 dye. After being tagged, the samples are hybridised to the slide, washed, scanned, and quantified in preparation for further computer analysis. In addition to being inexpensive, cDNA microarrays allow for the examination of two comparable biological samples at the same time. cDNA arrays can also assist in the discovery of new genes by detecting ESTs with unknown functions. They also provide the user the potential to add fresh cDNA clones and build smaller, tailored arrays for special inquiries. Spotted arrays have the drawback of being unable to determine absolute amounts of gene expression, which Affymetrix GeneChipsTM can do, as well as the fluctuation in spot quality from slide to slide. Their inability to adjust for nonspecific hybridisation, which can lead to incorrect gene expression estimations, is another problem.

Affymetrix (Santa Clara, CA, USA) manufactures short oligonucleotide microarrays, or GeneChipsTM (25 base pairs in length), using a photolithographic process similar to computer microchip creation. This method allows for the development of a high-density array with up to 1,000,000 distinct oligonucleotide characteristics spanning over 39,000 transcript variations. At least one set of 11–20 separate "probe pairs" represents each gene. A probe pair comprises a perfect-match (PM) and a mismatch (MM) pair, the latter of which is designed to not match the target sequence at the 13th position of the probe. In comparison to cDNA microarrays, the purpose is to adjust for nonspecific hybridisation and reduce noise in data interpretation.

A unique algorithm in the Affymetrix Microarray Suite software integrates data from all 20 paired PM and MM probes. Unlike cDNA arrays, which are hybridised with two samples, each mRNA preparation for an Affymetrix array is hybridised with a distinct Affymetrix GeneChipTM. The GeneChipTM is cleaned, stained, scanned, and quantified for further investigation after hybridisation. The advantage of Affymetrix GeneChipsTM is its widespread acceptance in the research community, which allows the user to benefit from an integrated platform that includes analytical tools in addition to the array itself. The biggest concern is their high cost and difficulty to compare the expression levels of two biological samples that are linked simultaneously (Kilpinen et al. 2008).

Microarray Data Analysis

The most critical requirement for a successful microarray experiment, as many researchers in the area have acknowledged, is proper experimental design. The most difficult component of a microarray process is dealing with the large amount of data generated by this technology and interpreting the results, which will entail the use of machine learning and statistical approaches.

The first step in a conventional microarray study is data preprocessing, which includes normalisation and filtering. Before comparing the expression data from the studies, this step is completed. In the computation of gene expression data, normalisation is required to account for and reduce systematic and experimental variances. It sought to find the biological information by reducing the impact of non-biological influences on the data.

Intensity-dependent normalisation and an intensity-independent technique are the two basic techniques. The number of genes, which often exceeds the number of observations by at least one order of magnitude, is the next obstacle to overcome after normalisation. Before any machine learning or statistical techniques can be used, significant variable reduction is generally required. Filtering the data, which minimises the number of genes before analysis, is one technique to solve this challenge. Genes with a low overall variation across all of the samples to be analysed, for example, can be filtered since they are of little interest (Alzubaidi et al. 2021).

Supervised or unsupervised machine learning methods are usually used to analyse microarray data. Unsupervised learning analyses data without making any assumptions about the identification of the samples being studied. Hierarchical cluster analysis, k-means clustering and self-organising maps are examples of unsupervised learning techniques used in microarray research. Unsupervised microarray data clustering has been widely employed in the categorisation of malignancies, such as leukaemia, breast, and prostate. In contrast, supervised learning approaches incorporate biological information for the samples being studied into the data analysis. This analysis is used to identify gene subsets that can predict a diagnosis or a clinical result. Nearest neighbour algorithms, linear discriminant analysis, classification trees, and support vector machines are examples of supervised learning approaches. Following data analysis, the selected genes must be biologically validated. A microarray experiment, as previously said, is a multistep procedure prone to mistakes, biases, and occasionally overinterpretation. Furthermore, data quality concerns will have a substantial impact on the final outcomes. At the RNA level, validation is usually done using one of three methods: Northern blot, real-time PCR, or in situ hybridisation on tissue sections.

Although microarray data processing remains difficult, numerous research organisations have since solved most of the original challenges. Despite the fact that this technique has become commonplace in many laboratories, an integrated microarray database is still needed to allow comparison of microarray data generated by different laboratories and microarray systems. Consistency and reproducibility across high-throughput technologies are major challenges since there are multiple microarray platforms for assessing gene expression. The two independent OSCC experiments mentioned in the previous section not only used two distinct platforms (cDNA and Affymetrix GeneChipsTM), but they were also done in two separate laboratories to demonstrate this problem. In principle, data collected by multiple laboratories and platforms should be able to be combined.

Transcriptomic Analysis

Transcriptomic analysis is a functional study at the genome level (large-scale DNA sequencing that revealed variation in coding sections of genes, resulting in gene expression polymorphisms). In recent years, significant progress has been made in understanding the genetic basis of oral cancer development. The development from a normal cell to a cancer cell is understood to be based on an accumulation of genetic abnormalities. The more abnormal function of genes that favourably or negatively affect elements of proliferation, apoptosis, genome integrity, angiogenesis, invasion, and metastasis facilitate progression. Several studies, however, imply that early genetic alterations are not always associated with altered morphology.

Many tumours have augmented gene dosage by DNA amplification, which leads to up-regulation of tumour-promoting genes. Much research contributed to existing knowledge; however, they did not explain the complexities of such cancers. Genomic-scale differential gene expression profiles may now be produced thanks to DNA microarrays. By connecting variations in gene expression patterns with particular medical disorders, microarray analysis has contributed to the discovery of knowledge that is valuable for therapeutic practice. A collection of samples that fall into one of two categories, such as pathological specimens versus healthy tissue controls or early stage versus late stage of a disease, is used to construct mRNA expression profiles for thousands of genes at once.

Large-scale microarray technology and/or high-throughput quantitative PCR can be used to examine the transcriptional features of tumours with unparalleled precision. Because genotypic changes precede phenotypic changes, it may be feasible to find differentially expressed genes that may be used as markers for the early discovery of premalignant lesions, as well as cancer invasion and metastasis.

Genomic-scale expression profiles allow researchers to examine genetic expression variations in proliferation, invasion, carcinogenic metabolism, apoptosis, and other genetic themes and pathways, rather than only focusing on the expression of a few genes. The composition of microarray gene expression profiles is extremely unpredictable and depends on the selection of patients used to determine the signature, according to a review of tumour profiling research. Furthermore, several researchers determined that the outcomes of such research are overoptimistic due to insufficient sample sizes and incorrect validation. This is not unexpected because many prognosticative genes have similar, if not identical, expression patterns across tumour samples, making them interchangeable without altering prediction accuracy. Head and Neck Squamous Cell Carcinomas. (HNSCCs, especially OSCCs), are complex heterogeneous tumours, and microarray technology enables for quick, large-scale gene screening, which might lead to the identification of candidate genes and pathways implicated in such a complicated illness.

The GeneChip, which is supplied by Affymetrix, is one of the "gold standard" technologies. The GeneChips are made utilising a mix of photolithography and solid-phase DNA synthesis processes. The probes on the Affymetrix GeneChip Microarray are 25 nucleotide bases long, and each gene has several oligonucleotide probes, resulting in great specificity when compared to other Microarray systems. Per GeneChip, just one mRNA population may be investigated, allowing for an absolute determination of gene expression level. Within the research community, the Affymetrix GeneChip Microarray platform has been generally accepted as a standard. The capacity to compare any single Microarray with any other and in any combination is a well-known benefit of the Affymetrix Microarray platform. Applied Biosystems, Beckman Coulter, Eppendorf Biochip Systems, Agilent, and Illumina are among the other companies that sell DNA microarrays.

Salivary Proteome Analysis

Top-down and bottom-up analytical methods for salivary proteomics are loosely categorised. To analyse salivary proteins that have been broken down, bottom-up proteomics is used. Bottom-up proteomics is used to analyse salivary proteins that have been digested, whereas top-down proteomics concentrates on the study of the intact naturally occurring proteome. Two-dimensional gel electrophoresis, a top-down platform, is the most fundamental method for sorting complicated protein mixtures including more than 5000 proteins. This approach, however, has some drawbacks: First, very abundant proteins may mask less abundant proteins; second, small proteins or peptides with a very acidic or basic isoelectric point (pI) may migrate outside of its analysis ranges; and third, this method has numerous drawbacks, including gel preparation, unusual migration, and staining of protein isoforms. The main methodology for proteomics is the use of mass spectrometry (MS) in conjunction with a variety of separation techniques. The "bottom-up" method, which analyses peptides through proteolytic digestion, and the "top-down" method, which examines intact proteins, are the two main methods used to identify and characterise proteins using MS. Salivary proteomes' levels of expression and posttranslational modifications can be examined using MS, which has a quick turnaround time and great sensitivity. To quantify intact proteins or peptides, this technique is typically paired with surface-enhanced laser desorption ionisation (SELDI), matrix-assisted laser desorption ionisation (MALDI), or time-of-flight (TOF). In the Protein

Chip, sample purification, desorption/ionisation, and protein separation are all crucially impacted by SELDI–TOF–MS. In contrast, MALDI–TOF–MS can be utilised for preliminary profiling before further identification by HPLC–MS because of its superior sensitivity and simplicity. Absolute protein quantification using internal standards (AQUA), stable isotope labelling by amino acids in cell culture, isotope-coded affinity tags, isotope tags for relative and absolute quantification, AQUA, and other chemical labelling techniques are also frequently used to observe quantitative changes of the salivary proteome in transitional stages (Pearce & Zhang 2021).

Bottom-Up Analysis

Proteins are digested by proteases before being analysed by MS in bottom-up proteomics (Figure 1.2). Bottom-up refers to the process of reconstructing information about the individual proteins from each detected fragment peptide. According to the bottom-up proteomics flowchart; two-dimensional electrophoresis is used to separate the protein mixture using the gel method. Proteins are collected from the gel, digested, and analysed by MS after spot visualisation to be further identified by database searching. Utilising a gel-free procedure, the protein mixture is immediately digested into a mixture of peptides, and the peptides are subsequently separated using multidimensional separation techniques. MS is then used to analyse the peptides. Database searching is used to identify proteins from the generated mass spectra. The most developed and popular method for identifying and characterising proteins is bottom-up proteomics. It does not require sophisticated equipment or special knowledge. Additionally, it has the capacity to separate materials with great resolution. Bottom-up proteomics has a lot of restrictions. Only a limited and unstable portion of

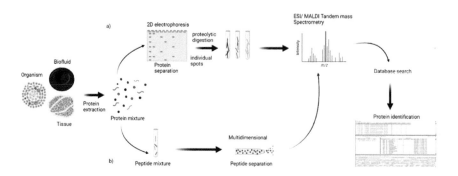

FIGURE 1.2   The flowchart of bottom-up proteomics.

a protein's entire peptide population may be retrieved, resulting in low percentage coverage of the protein sequence. According to genomic research, each open reading frame can result in a number of protein isoforms due to alternative splicing products and variations in posttranslational modification (PTM) sites and types. Due to the low sequence coverage in bottom-up proteomics, a lot of information on PTMs and alternative splice variants is lost.

Top-Down Proteomics

Top-down characterising intact proteins from intricate biological systems is possible with proteomics. As a result of a combination of genetic diversity, alternative splicing, and post-translational changes, forms the precise molecular shape of the protein can be fully characterised and covered by this method almost entirely. Either electron-capture dissociation or electron-transfer dissociation (ETD) is used to achieve fragmentation for tandem MS. In a Fourier transform ion cyclotron resonance or quadrupole ion trap mass spectrometer, proteins are typically electrospray ionised and trapped. Without first digesting the proteins in the complex mixtures into their appropriate peptide species, it requires protein identification. Protein separation, mass spectrometer detection, and data analysis are included in the top-down proteomics workflow diagram (Figure 1.3). The possible access to the entire protein sequence and the capacity to identify and describe PTMs are the key benefits of the top-down approach. Additionally, it has the capacity to identify protein isoforms. Additionally, the bottom-up approaches' time-consuming protein digestion is dropped. Although there are some drawbacks, the high costs are a major factor in

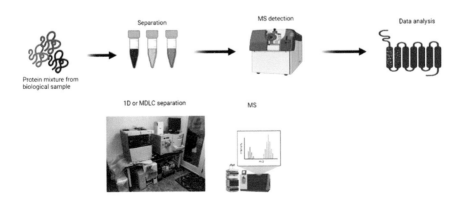

FIGURE 1.3   The flowchart of Top-Down Proteomics.

the widespread adoption of top-down proteomics. The preferred instrumentation is costly to buy and maintain. Additionally, the preferred dissociation methods (ECD, ETD) are low-efficiency procedures that demand lengthy ion buildup, activation, and detection durations. Due to a lack of intact protein fractionation techniques that are coupled with tandem MS, it has not been accomplished on a wide scale. Through the use of peptides produced by the proteolytic digestion of intact proteins, the bottom-up method offers an indirect way to measure proteins. Prior to being introduced into the mass spectrometer, whole proteins are digested into peptides, where they are then identified and fragmented. The top-down approach, in contrast, assesses the intact proteins Pearson 2013a and 2013b. For a thorough picture of all proteoforms, including those with PTMs and sequence variants, the protein is isolated from cell or tissue lysates, separated by gel or LC, and then directly analysed by MS (Figure 1.4).

Selection of Bioinformatic Approaches

The need to maintain and evaluate the growing volume of biotechnological datasets led to the development of the specialised scientific field known as bioinformatics. It can be broadly characterised as the application of machine learning and statistical approaches to analyse large-scale, multidimensional data collected from many sources to research biological processes. It involves setting up data storage, creating tools for data analysis, and actually analysing the data. To extract analyse data, bioinformatics employs both informatics and statistical methods.

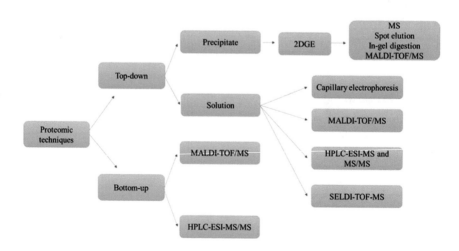

FIGURE 1.4   Integrated top-down and bottom-up proteomics approaches.

Cancer biology is frequently connected to informatics as a result of the rapid development of large-scale databases by high throughput research platforms, including gene expression assays. Bioinformatics is mostly used to discover cancer biomarkers, their functions, and the molecular pathways that drive cancer growth. A large number of unique bioinformatics techniques have already been identified at a sophisticated level, allowing for semi-automated functional analysis, such as gene set enrichment and pathway-based analysis. A wide range of bioinformatics techniques may be used to undertake functional analysis of biomarkers discovered in large-scale datasets. Gene annotation is the most important step in starting a functional analysis.

A wide range of bioinformatics tools can be used to perform functional analysis of biomarkers discovered in large-scale datasets. Gene annotation is the most important step in starting a functional analysis. Various biotechnological database management organisations, such as NCBI (Gene), EBI (Ensembl), and caArray database, as well as independent groups, like GeneCards provide annotative information. Kyoto Encyclopedia of Genes and Genomes (KEGG) and BioCarta are two pathway information systems that organise information about the molecular interactions of gene products. Another way to improve functional analysis is to associate genes with their corresponding ontological terms. The gene ontology (GO) database was created to standardise the attributes of genes and gene products across a vast array of biological databases. DAVID (Database for Annotation, Visualization, and Integrated discovery) is a hybrid functional analysis tool that brings all of these systems together in one web interface. Such integrative tools improved gene and protein functional analysis while also providing a holistic view of large-scale datasets.

Microarray Analysis

Microarray technologies have become a popular tool for measuring gene expression profiles in research. This method can be used to simultaneously examine thousands of genes. The DNA microarray is a technique for determining whether or not a person's DNA contains a gene mutation. A small glass plate is encased in plastic to make up the chip. Thousands of short, synthetic, single-stranded nucleotide sequences complementary to the normal gene or variants of that gene are contained on a microarray chip. Bioconductor, an open-source and open-development software project based on the R programming language, is one of the most popular platforms for microarray analyses. Arrays can be used to classify and identify gene profiles linked to a variety of cancers, including oral cancer. This kind of genetic approach will help researchers better understand disease progression, which will lead to better diagnosis and treatment for patients. Based on their expression profile, it can help identify and classify genes linked to cancers, periodontal diseases, and caries. The use of bioinformatics analysis in the processing of data obtained from microarray analysis is critical.

The variety of severities, durations, sensitivity and resistance to drugs, cell differentiation, and origin, and understanding of pathogenesis play a role in disease diagnosis, therapies, and prognoses. The majority of the time, cancer is discovered in its later stages, reducing the chances of cure or improvement. Early detection is now possible thanks to advances in Artificial Neural Networks (ANNs) and image analysis of biopsy samples. The brain is made up of billions of interconnected neurons that can handle complex and computationally intensive tasks. ANN was created to mimic the learning ability of biological neurons. ANN can be used in predictive or diagnostic models because of its learning ability and computational power. Because there is growing evidence that gene–protein interactions are important in understanding cancer molecular mechanisms, it is necessary to study cancer at the molecular level. The role of bioinformatics in understanding the mechanism and effects of treatment is critical as the healthcare industry moves toward personalised medicines. Information about a patient's genetic, enzymatic, and metabolic profile is used to develop personalised medicine that is tailored to that person's needs and to assess a patient's risk factors.

Bioconductor

Bioconductor is bioinformatics open-source software that includes tools for analyzing and comprehending high-throughput genomic data. Affymetrix, Illumina, NimbleGen, Agilent, and other one- and two-colour technologies are all supported by Bioconductor's advanced facilities for microarray analysis. Exon, copy number, SNP, methylation, and other assays are well-supported in Bioconductor, as is extensive support for expression array analysis. Pre-processing, quality assessment, differential expression, clustering and classification, gene set enrichment analysis, and genetical genomics are some of the major workflows in Bioconductor. GEO, ArrayExpress, BioMart, genome browsers, GO, KEGG, and a variety of annotation sources are just a few of the community resources available through Bioconductor.

Data Filtering and Normalisation

Normalisation is required to explain and reduce systematic and experimental variations in the calculation of gene expression data, allowing for the identification of biologically relevant data and the removal of non-biological effects. The Gene Chip Robust Multichip Average method (gcRMA) was the preferred method in the thesis research. Before using machine learning or statistical algorithms, the data is filtered to reduce variation after normalisation. Because they are of little value, genes with a low overall variance across all samples can be removed. Many analysis packages include filtration as part of the first stage of processing.

Corrections of Batch Effects

Researchers can use microarray datasets to increase statistical power in detecting biological phenomena in studies where sample size is limited due to logistical constraints. In general, combining datasets without accounting for batch effects is not a good idea. Laboratory conditions, reagents, and personnel diversity can all cause batch effects. This becomes a major issue when batch effects are linked to a biologically relevant outcome and lead to incorrect conclusions.

To minimise effects between studies, the ComBat batch correction method was used in this study. For adjusting data of batch effects, the ComBat package in R software used parametric and nonparametric empirical Bayes frameworks. The best performance came from ComBat IS.

Differential Gene Expression Analysis

To examine differential gene expression and extract meaningful biological results from this comparison, samples are compared relative to one another. In general, there are two types of data analysis. The first is "unsupervised" analysis, which involves sorting data and calculating gene expression differences based on inherent differences in the dataset. Using Pearson correlation and hierarchical clustering in R, unsupervised clustering is achieved and heatmaps are generated. The Significance Analysis for Microarrays (SAM) in the siggenes package from Bioconductor is used for supervised analysis. SAM is one of the more powerful and effective methods for identifying specific gene products that play a role in distinguishing transcriptional profiles between two groups (Selvaraj & Natarajan 2011).

Biological Annotations of Significant Genes and Metabolites

To define the relative cellular functions, canonical pathways, upstream regulators, potential biomarkers, and molecular processes that are most frequently perturbed in correlation with a set of altered genes or metabolites of a particular comparison, as well as to import differentially expressed genes and/or metabolites from various comparison analyses, they are then uploaded and analysed using IPA tools. The IPA knowledge base enlists the well-known genes (and/or metabolites) that are involved in and control a certain disease or ailment to generate and emphasise the aforementioned functions. The *p*-value (right-tailed Fisher's exact test) determines the significance of a canonical pathway. This strategy has the benefit of creating routes without using the conventional canonical pathways database.

As a result, this technique may make linkages and correlations between genes and their by-products, such as metabolites, that were previously unknown. At least one reference from the literature corresponds to a line showing the connections between the genes (the pathway's nodes). The significance of a canonical pathway is assessed using the *p*-value (right-tailed Fisher's exact test).

Gene expression analysis undoubtedly benefits from the use of pathway analysis to further evaluate candidate genes and their potential role in each disease, while the discovery of regulatory networks sheds light on the pathophysiology of a particular disease (Rahman et al. 2016).

Patient Care Databases

Patient care databases are online repositories of information about a patient's diagnosis, procedures, and drug prescriptions, among other things. Rare case information can be stored and accessed quickly. Machine learning algorithms are applied to the collected data and can be used to provide appropriate and cost-effective treatment. To collect up-to-date information from a visiting patient, a personalised, well-designed hospital database can be used. This allows you to access relevant information with a single click of a button. If a patient requires the services of healthcare providers from different hospitals, information can be exchanged. It is a crucial tool for tracking and improving healthcare services. It can also help with billing and documentation. Medical facility operating costs can be reduced by reducing paperwork and clerical staff.

Genome-Wide Analyses – From Genome to Proteome

DNA sequencing is critical to the progress of molecular biology, not only because it changes the landscape of genome designs but also because it opens new prospects and applications. There are several applications of NGS technologies.

Genome

Because of the lower cost of sequencing, several genomes have been published. The size and quality of the readings (150–300 bp) are a constraint of the current techniques, which poses a difficulty for assembly software. They do, however, create a lot more sequences.

The genome must be assembled by making sense of the millions of base pairs sequenced. The assembly is made up of a hierarchical data structure that connects the sequence data to a target reconstruction. When a genome is sequenced, there are two options: If the genome of the species has already been constructed (reference genome), mapping with the reference genome is performed. However, if a genome has never been defined before (de novo), assembly is necessary (Trevino et al. 2007).

Sequencing data is stored in the sequencer as luminance pictures collected during DNA synthesis. As a result, the calling base refers to the process of acquiring picture data and converting it to a DNA sequence (FASTA). Also acquired is the value of each base's quality, known as the Phred quality score. Quality control is the process of evaluating the quality of sequenced reads (Phred value) and filtering out low-quality bases and adaptor sequences. To generate continuous fragments that correspond to the overlap of two or more reads, each of the reads is mapped to each other in the search for identity or overlapping areas.

The order, direction, and widths of gaps between contigs are defined by super contigs, also known as scaffolds. Overlap/Layout/Consensus or Bruijn's graphs techniques can be used to find the overlay regions. Only seeds that share reads are examined in these graphs, which use a technique based on seed alignment.

The sequenced reads and their overlaps are represented by the overlapping graph, which must be generated beforehand using a pairwise series of alignments. The readings are represented by the nodes, while the overlays are represented by the edges. The reading arrangement and contig consensus sequence are then computed using the overlapping graph. Bruijn's graph, on the other hand, minimises computing work by dividing reads into tiny DNA sequences known as k-mers. The length of the sequence's bases, which are always overlaid $k-1$ between k-mers, is denoted by the parameter k. Some indices, such as coverage, which refers to the number of reads associated with a specific DNA fragment, are used to assess the assembly quality. The N50 indicates the percentage of the genome covered by large contigs. An N50 value of n indicates that 50% of reads are contained in contigs of size n or greater.

The annotation of the genome, which is the extraction of biological information contained in the sequences, is the next stage in the genome assembly process. Because of the variations between prokaryotes and eukaryotes, several methodologies for searching for genes in genomes have been created. The approach begins with identifying genes based on sequence similarity. Following that, the gene function is annotated using protein databases such as NCBI and UniProt. Functional annotation, which involves using GO words to link genes to biological processes, is also performed. Gene function is classified into three categories: Molecular function, biological processes, and cellular components.

Transcriptomics

To infer and measure the transcriptome, DNA sequencing or hybridisation methods have been created. Although real-time PCR (qPCR) and DNA microarray have enabled significant progress, they have limits (Marioni et al. 2008; Wang et al. 2009). NGS platforms, on the other hand, have emerged as a viable alternative to these technologies for assessing global expression.

When compared to other technologies, RNA-seq allows mapping reads and transcript-level quantification with high-throughput, quantitatively and more precisely, and at a cheaper cost. Differential expression analysis and identification of isoforms arising from splicing, as well as the discovery of novel transcripts, such as long non-coding RNAs (lncRNAs), microRNAs, and allele-specific expression, are some of the other uses of RNA-seq. All of these options have allowed us to learn more about the genome's organisation, the molecular constituents of cells and tissues, and the complexities of regulatory systems. The differential expression technique has been emphasised among the mRNA analysis approaches. It is possible to discover genes whose abundance has changed dramatically between experimental settings using this method.

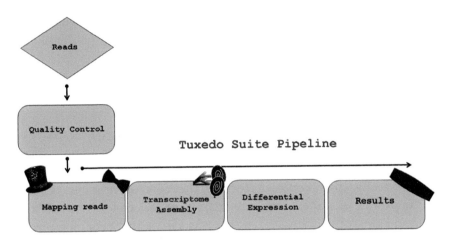

FIGURE 1.5   Tuxedo suite protocol for differential expression analysis.

The mRNA for the conditions to be evaluated must be extracted, purified, split into small pieces, and reverse transcriptase transcribed to cDNA to form an RNA-seq dataset. After that, the adapters are joined, and the pieces are sorted by size. Finally, the cDNAs are read using NGS.

There is a myriad of data analysis tools accessible. These analyses can be categorised into three categories: (1) read mapping; (2) transcript assembly; and (3) gene/transcript quantification. One of the most extensively used tools is the Tuxedo Suite protocol (Figure 1.5).

In general, these analyses are carried out in the following manner: Tophat finds splicing junctions and conducts read mapping to the reference genome. These alignments are then employed by the Cufflinks, which assemble the transcripts, assess their abundance, and calculate the genes and differentially expressed transcripts (Cuffdiff) under the circumstances studied, which CummeRbund can see. It is feasible to do functional enrichment analysis and biological process annotation using the list of genes acquired from differential expression analysis. In the KEGG and Reactome databases, for example, it is possible to find biological pathways in which these genes participate.

Proteomics

Understanding the molecular mechanisms that mediate cellular physiology requires the identification, measurement, and characterisation of all cell proteins. Proteomics appears to have exploded in popularity in the quest to organise the study of protein structure, function, relationships, and dynamics in space and time.

Protein identification can be accomplished in three ways: Direct protein sequencing, electrophoresis gels, and MS. The MS has revolutionised proteomics by providing both measurement of expression and characterisation of posttranslational modifications by detecting proteins in complex mixtures with high sensitivity. A new name "Next-generation proteomics" was coined to describe this technology.

A mass spectrometer is used to determine the proteome's composition. This consists of (1) an electrospray ionisation or matrix-assisted laser desorption/ionisation system; (2) one or more Time-of-Flight (TOF; Ion Trap) systems; and (3) one detector. The first component is developed to create peptide or protein ions, which are then accelerated by an electric field and separated by mass/charge (m/z) in a mass analyser or are chosen according to a specified m/z and fragmented in a tandem (MS/MS) process. Finally, the ions are pumped through a detector that is linked to a computer with data analysis software.

The procedures required to determine the proteome, according to, are as follows. The initial step is to extract proteins from the target tissue, which is then digested using a protease to produce peptides. These are fractionated to minimise the sample's complexity, using procedures like liquid chromatography, or enriched, identifying subsets of the sample with affinity for resins or antibody immunoprecipitation (sample preparation). The intact peptide's m/z is recorded by the spectrometer after ionisation. The most abundant peptides are then chosen, collision fragmented, and subjected to the tandem MS/MS procedure. The ions y and b are produced because of this action, and they are equivalent to segments of the C- and N-terminal regions, respectively.

The generated spectrum is a collection of m/z ratios for different fragments, each of which has a mass difference equal to a specific amino acid. After that, the list is compared to MS databases like MASCOT. Finally, the data is quantified for the various experimental circumstances and proteins, and the results are interpreted in light of the biological topic being investigated.

The found proteins may be linked to GO terms and used to build biological pathways. Another alternative is to study protein–protein interactions using co-expression or databases like MINT and BioGRID.

Biological Interpretation of the Common Expressed Genes

To better understand the biological reasons why the expressed genes in OSCCs differ from those in normal oral tissue samples, the Ingenuity pathway analysis was used to look at the molecular function, biological process, canonical pathways, and upstream/transcriptional regulators analysis of the genes. The IPA system generates an integrated graphical representation of the biological relationships between genes and gene products, taking into account both up- and downregulated genes in the comparison analysis. It does this by using a knowledge database of selected functional and regulatory interactions extracted from the literature.

Functional Analysis

The top of the perturbed molecular cellular functions in relation to cancer is displayed as cell trafficking, movement, growth, proliferation, development, death, and survival.

Associated Network Analysis

Using a network analysis of common genes, researchers looked into the biological relationships between genes and gene products. The highly connected nodes are likely to reflect their ability to control a large number of genes in the master network, as well as the gene expression pattern identified in the comparison's tumour signature. At the centre of the network, overexpression of VEGFA, HIF1A, TGFB1, IL8, IL1B, FN1, SPP1, STAT1, SERPINE1, MMP1, MMP3, MMP7, MMP9, MMP13, TNC, RHOA, PLAU, CXCL10, CXCL2, THBS1, COL1A1, COL1A2, COL3A1, and PTHLH was found, as well as underexpression.

Upstream Regulators

It is common knowledge that the regulation of human genes is dependent on a group of transcription factors rather than a single factor. As a result, the IPA system's "Upstream and Transcriptional Regulator Analysis" tool was used to identify several key players in the biological context. CTNNB1, JUN, NFKBIA, and STAT3 are activated, while TP53 and MYC transcriptional regulators are inhibited. Activation of TGFB1, EGF, and EGFR growth factors and their receptors was also associated with activation of TNF and IFNG cytokines, as well as repression of the ERBB2 cytokine.

Canonical Pathway Analysis

Role of tissue factor in cancer, leukocyte extravasation signaling, inhibition of matrix metalloproteases, HIF1a signaling, ILK signaling, and xenobiotic metabolism signaling were identified as the most commonly perturbed pathways involved in tumour transformation using the canonical pathway analysis tool (IPA). The role of these pathways during tumour transformation has been extensively described in previous research. By overlapping the input common genes over the pre-generated pathways, the significant dysregulated pathways can be identified.

CONCLUSION

In conclusion, research into cancer, a major cause of death worldwide, is still very difficult. Effective early diagnosis is therefore crucial, but it can be difficult with noncomputational procedures because they are more expensive, less reliable, and some, like CT scans, expose people to radiation. As a result, bioinformatics has emerged as a very effective and ground-breaking method for enhancing cancer diagnosis.

REFERENCES

Abdurakhmonov IY. Bioinformatics: Basics, Development, and Future. In: Abdurakhmonov IY, editor. Bioinformatics – Updated Features and Applications. London: IntechOpen. 2016. Available from: https://www.intechopen.com/chapters/50934

Alzubaidi L, Zhang J, Humaidi AJ et al. Review of deep learning: concepts, CNN architectures, challenges, applications, future directions. J Big Data. 8, 53 (2021). https://doi.org/10.1186/s40537-021-00444-8.

Bayat A. Science, medicine, and the future: Bioinformatics. BMJ. 2002;324(7344): 1018–1022. doi: 10.1136/bmj.324.7344.1018.

Berger MF, Mardis ER. The emerging clinical relevance of genomics in cancer medicine. Nat Rev Clin Oncol. 2018;15(6):353–365. doi: 10.1038/s41571-018-0002-6.

Borse V, Konwar AN, Buragohain P. Oral cancer diagnosis and perspectives in India. Sens Int. 2020;1:100046. doi: 10.1016/j.sintl.2020.100046.

Brem H, & Tomic-Canic, M. (2007). Cellular and molecular basis of wound healing in diabetes. The Journal of clinical investigation. 117(5):1219–1222. https://doi.org/10.1172/JCI32169

D'Afonseca V, Prosdocimi F, Dorella FA, Pacheco LG, Moraes PM, Pena I, Ortega JM, Teixeira S, Oliveira SC, Coser EM, Oliveira LM, Corrêa de Oliveira G, Meyer R, Miyoshi A, & Azevedo V (2010). Survey of genome organization and gene content of Corynebacterium pseudotuberculosis. Microbiological research, 165(4):312–320. https://doi.org/10.1016/j.micres.2009.05.009

Dalal A, & Atri A (2014). An Introduction to Sequence and Series. International Journal of Research, 1(10):1286–1292.

Fiser A. Template-based protein structure modeling. Methods Mol Biol. 2010;673:73–94. doi: 10.1007/978-1-60761-842-3_6.

Garcia I, Kuska R, Somerman M. (2013). Expanding the foundation for personalized medicine: Implications and challenges for dentistry. J Dent Res. 92(7 Suppl):3S–10S. doi: 10.1177/0022034513487209.

Gauthier J, Vincent AT, Charette SJ, Derome N. A brief history of bioinformatics. Brief Bioinform. 2019;20(6):1981–1996. doi: 10.1093/bib/bby063.

Gupta N, & Verma VK (2019). Next-Generation Sequencing and Its Application: Empowering in Public Health Beyond Reality. Microbial Technology for the Welfare of Society, 17:313–341. https://doi.org/10.1007/978-981-13-8844-6_15

Kilpinen S, Autio R, Ojala K, Iljin K, Bucher E, Sara H, Pisto T, Saarela M, Skotheim RI, Björkman M, Mpindi JP, Haapa-Paananen S, Vainio P, Edgren H, Wolf M, Astola J, Nees M, Hautaniemi S, & Kallioniemi O (2008). Systematic bioinformatic analysis of expression levels of 17,330 human genes across 9,783 samples from 175 types of healthy and pathological tissues. Genome biology, 9(9), R139. https://doi.org/10.1186/gb-2008-9-9-r139

Kuo WP. Overview of bioinformatics and its application to oral genomics. Adv Dent Res. 2003;17:89–94. doi: 10.1177/154407370301700121.

Manzoni C, Kia DA, Vandrovcova J, Hardy J, Wood NW, Lewis PA, Ferrari R. Genome, transcriptome and proteome: The rise of omics data and their integration in biomedical sciences. Brief Bioinform. 2018;19(2):286–302. doi: 10.1093/bib/bbw114.

Marioni JC, Mason CE, Mane SM, Stephens M, & Gilad Y (2008). RNA-seq: an assessment of technical reproducibility and comparison with gene expression arrays. Genome research. 18(9):1509–1517. https://doi.org/10.1101/gr.079558.108

Maria Chatzou, Cedrik Magis, Jia-Ming Chang, Carsten Kemena, Giovanni Bussotti, Ionas Erb, Cedric Notredame, Multiple sequence alignment modeling: methods and applications, Briefings in Bioinformatics, Volume 17, Issue 6, November 2016, Pages 1009–1023, https://doi.org/10.1093/bib/bbv099

Oulas A, Minadakis G, Zachariou M, Sokratous K, Bourdakou MM, Spyrou GM. Systems bioinformatics: Increasing precision of computational diagnostics and therapeutics through network-based approaches. Brief Bioinform. 2019;20(3):806–824. doi: 10.1093/bib/bbx151.

Pearce R, Zhang Y. Deep learning techniques have significantly impacted protein structure prediction and protein design. Curr Opin Struct Biol. 2021;68: 194–207. doi: 10.1016/j.sbi.2021.01.007.

Pearson WR. Selecting the right similarity-scoring matrix. Curr Protoc Bioinformatics. 2013a;43:3.5.1–3.5.9. doi: 10.1002/0471250953.bi0305s43.

Pearson WR. An introduction to sequence similarity ("homology") searching. Curr Protoc Bioinformatics. 2013b;Chapter 3:Unit3.1. doi: 10.1002/0471250953.bi0301s42.

Pires FR, Ramos AB, Oliveira JB, Tavares AS, Luz PS, Santos TC. Oral squamous cell carcinoma: Clinicopathological features from 346 cases from a single oral pathology service during an 8-year period. J Appl Oral Sci. 2013;21(5):460–467. doi: 10.1590/1679-775720130317.

Piyarathne NS, Rasnayake RMSGK, Angammana R, Chandrasekera P, Ramachandra S, Weerasekera M, Yasawardene S, Abu-Eid R, Jayasinghe JAP, Gupta E. Diagnostic salivary biomarkers in oral cancer and oral potentially malignant disorders and their relationships to risk factors – A systematic review. Expert Rev Mol Diagn. 2021;21(8):789–807. doi: 10.1080/14737159.2021.1944106.

Rahman M, Shaheen T, Rahman M, Iqbal MA, Zafar Y. Bioinformatics: A Way Forward to Explore "Plant Omics". In: Abdurakhmonov IY, editor. Bioinformatics - Updated Features and Applications [Internet]. London: IntechOpen. 2016. Available from: https://www.intechopen.com/chapters/51698 doi: 10.5772/64043.

Selvaraj S, Natarajan J. Microarray data analysis and mining tools. Bioinformation. 2011;6(3):95–9. doi: 10.6026/97320630006095.

Singaraju S, Prasad H, Singaraju M. Evolution of dental informatics as a major research tool in oral pathology. J Oral Maxillofac Pathol. 2012;16(1):83–87. doi: 10.4103/0973-029X.92979.

Trevino V, Falciani F, Barrera-Saldaña HA. DNA microarrays: A powerful genomic tool for biomedical and clinical research. Mol Med. 2007; 13(9–10):527–541. doi: 10.2119/2006-00107.Trevino.

Vodkin LO, Khanna A, Shealy R, Clough SJ, Gonzalez DO, Philip R, Zabala G, Thibaud-Nissen F, Sidarous M, Strömvik MV, Shoop E, Schmidt C, Retzel E, Erpelding J, Shoemaker RC, Rodriguez-Huete AM, Polacco JC, Coryell V, Keim P, Gong G, Liu L, Pardinas J, Schweitzer P. Microarrays for global expression constructed with a low redundancy set of 27,500 sequenced cDNAs representing an array of developmental stages and physiological conditions of the soybean plant. BMC Genomics. 2004;5:73. doi: 10.1186/1471-2164-5-73.

Wang Y, Wu H, Cai Y. A benchmark study of sequence alignment methods for protein clustering. BMC Bioinformatics 19 (Suppl 19), 529 (2018). https://doi.org/10.1186/s12859-018-2524-4

Wang Z, Gerstein M, & Snyder M (2009). RNA-Seq: a revolutionary tool for transcriptomics. Nature reviews. Genetics, 10(1), 57–63. https://doi.org/10.1038/nrg2484

Xu D, Xu Y. Protein databases on the internet. CurrProtoc Mol Biol. 2004;Chapter 19:Unit 19.4. doi: 10.1002/0471142727.mb1904s68.

Yakob M, Fuentes L, Wang MB, Abemayor E, Wong DT. Salivary biomarkers for detection of oral squamous cell carcinoma – Current state and recent advances. Curr Oral Health Rep. 2014;1(2):133–141. doi: 10.1007/s40496-014-0014-y.

CHAPTER 2

Background of Oral Cancer

Viveka S[1], Bhargav Shreevatsa[2+], Kavana C P[2],
Anisha S Jain[3] Chandan Dharmashekara[2],
Bhavana H H[3], Sumitha. E[2], Mahesh KP[1],
Shiva Prasad Kollur[4], Chandan Shivamallu[2]

[1]Department of Oral Medicine and Radiology,
JSS Dental College and Hospital, JSS Academy of Higher
Education & Research, Mysuru, Karnataka, India

[2]Department of Biotechnology and Bioinformatics, JSS Academy
of Higher Education and Research, Mysuru, Karnataka, India

[3]Department of Microbiology, JSS Academy of Higher
Education and Research, Mysuru, Karnataka, India

[4]School of Physical Sciences, Amrita Vishwa
Vidyapeetham, Mysuru, Karnataka, India

INTRODUCTION

Any malignant development in the mouth is referred to as oral cancer (OC), which is a subgroup of head and neck malignancies. This malignancy can be deadly if left untreated and comprises tumors of the tongue, lips, cheeks, gum, floor of the mouth, soft and hard palate, tonsils, sinuses, and throat. In the squamous cells that line the surface of the mouth, more than 90% of different kinds of OC begin. OC has an incidence of four cases per 100,000 people worldwide (age-adjusted), with significant regional differences depending on gender, race, countries,

DOI: 10.1201/9781032625713-2

age groups and ethnic groups, and socioeconomic conditions (World Health Organization, 2020). It is the 16th most common malignancy and the 15th leading cause of global death. Unquestionably, the majority of the disparities between the developed and developing world stem from the various population habits, preventive education, life expectancies, and medical record quality in each region (Ferlay et al. 2019; García-Martín et al. 2019). Oral cavity cancer has several physical characteristics, environmental variables, and hereditary risk factors.

In the present, while many physical conditions and environmental agents are established risk factors (Panta 2019) for OC, they can be effortlessly fought with prevention commercials, and oral surgeons must look into and understand more about those conditions whose major role on OC pathogenesis is still unknown and mysterious. There have been several risk factors or potential causes of OC reported. It has been demonstrated that OC is greatly influenced by chemical components like alcohol and cigarette use, biological ones like the human papillomavirus (HPV), syphilis, oro-dental factors, nutritional deficits, and viruses (Figure 2.1).

FIGURE 2.1   Basic causes of oral cancer.

Etiology of Oral Cancer

1. Smoking

2. Alcohol

3. Exposure to sun

4. Diet and nutrition

5. Viral infections

6. Genetics

7. Chronic irritation

On the basis of a basic visual examination, medical history, and an investigation of risk factors, OC can be detected. The survival rate of patients can be increased by up to 90% with early diagnosis and quick referral to specialized facilities. As a result of metastases and advanced stage III or IV diagnosis in roughly 60% of OC cases, the death rate is unfortunately greater (Wyss et al. 2013). Both patients' and healthcare providers' ignorance might be at blame for diagnostic delays. As oral malignancies are typically asymptomatic, the patients' delaying causes include late recognition of the lesion or symptoms, ignoring the lesions, self-medication, surgical anxiety, low socioeconomic situations, and limited or no access to specialist healthcare. Professionally, the problems might include incorrect intraoral and extraoral inspection, a delay in conducting the biopsy or choosing the inappropriate sample location for histopathological analysis. According to reports, it takes an average of six months between the initial diagnosis and the final diagnosis (Ram et al. 2011).

Educational programs that are population-targeted should primarily target high-risk populations. On the other hand, professional intervention strategies should provide a thorough understanding of the clinical presentation, particularly at locations like the gingivae, floor of the mouth, and retromolar trigone. Every nation should conduct screening initiatives at the primary or secondary care levels (Figure 2.2).

Molecular Pathogenesis of Oral Cancer

Like other cancers, oral carcinogenesis develops over time, with normal epithelium going through stages of dysplasia until becoming aggressive

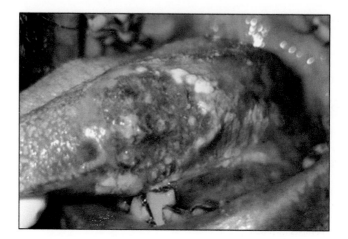

FIGURE 2.2   Tongue cancer.

variants. Squamous cell carcinoma (SCC) is the most prevalent variety of OC, despite the fact that all forms of carcinomas can be found in the oral cavity. In recent years, the molecular histopathologic picture of OC has been unveiled through the use of genomic and proteomic strategies. There is ongoing research to determine the role of epigenetic changes, genomic instability, and the generation of a gene expression profile in the development of OC (Jain 2019). Understanding these genetic modifications and the sequences of gene expression is essential for comprehending the molecular etiology of OC (Sujir et al. 2019). Despite some important advances, it will take another 10 years of intense study to fully grasp the molecular pathophysiology of OC and its relationship to the causal agent (Figure 2.3).

Genetic Susceptibility

Roughly 10% of all malignancies are now known to have a significant genetic component. Numerous studies that demonstrate familial clustering imply a biological basis in the evolution of OC. For head and neck cancer, including OC, the Glutathione S-transferase M1 null genotype tends as the most reliable polymorphism susceptibility marker. According to a meta-analysis of 12 research, the variation val allele of the CYP1A1 (Cytochrome-P450, family-1, member A1) polymorphism would be another reasonably reliable susceptibility marker. Numerous additional gene polymorphisms have not yielded significant results from investigations. Aldehyde dehydrogenase genes ALDH1B and ALDH2

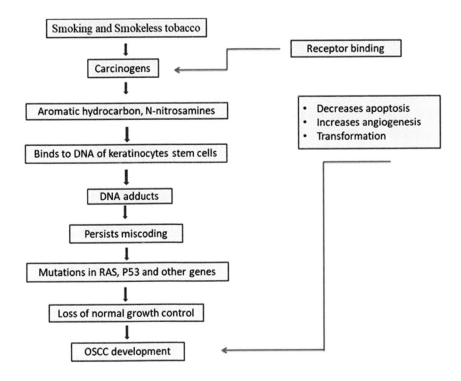

FIGURE 2.3 Molecular pathogenesis of OSCC. *OSCC*, Oral squamous cell carcinoma.

(Aldehyde dehydrogenase 2) were linked to Head and neck squamous cell carcinomas (HNSCC) and had a substantial connection with alcohol usage, according to Brennan et al. (2004).

Proto-Oncogenes, Oncogenes, and Genetic Alterations

Point mutations, amplifications, rearrangements, and deletions are examples of genetic abnormalities that characterize the molecular foundation of carcinogenesis. Oral carcinogenesis has also been linked to a number of oncogenes (Williams 2000). EGFR (epidermal growth factor receptor), K-ras, parathyroid adenomatosis-1, c-myc, int-2, and B-cell lymphoma-like oncogenes have all been documented to express themselves abnormally during the development of OC.

These three metabolic pathways are how proto-oncogenes function (Williams 2000):

The first method uses serine, threonine, and tyrosine as substrates to phosphorylate proteins. The second way that gene function is through the GTPase-mediated signaling of signals. The revelation of RAS oncogenes, which encode a previously unidentified form of GTPase, led to the initial understanding of the significance of these signaling mechanisms in cancer. DNA-based transcription regulation is the third method. Proto-oncogenes are still encoding a wide range of transcription factors that may also take part in DNA replication directly (Gudiseva et al. 2017).

Tumor Suppressor Genes

A p53 mutation is present in more than 50% of all primary HNSCC. The most frequent genetic alteration in all human malignancies is p53 inactivation. Chromosome 9p21–22 contains the HNC region that is most frequently deleted (Rivlin et al. 2011). The majority of invasive tumors in head and neck cancer lose chromosomal 9p21. The most prevalent genetic mutation seen in this area is the homozygous deletion, which is widespread. This deleted area contains p16 (CDKN2), a powerful cyclin D1 inhibitor. Additionally, loss of p16 protein has been seen in the majority of advanced premalignant lesions (Denaro et al. 2011). Most human cancers, including OC, frequently lose chromosome 17p. Nearly 60% of invasive lesions exhibit it. Even though p53 inactivation and loss of 17p are strongly correlated in invasive lesions, p53 mutations are extremely uncommon in early lesions with 17p loss. Primary cancers have also been seen to lose chromosomal arms 10 and 13q.

In OC, several other chromosomal loss areas have been observed. Additionally, detailed mapping of these crucial genes inside these regions may offer crucial details for understanding the genomic instability that contributes to the formation of this neoplasm.

Growth Factor Receptors and Mechanisms

Increased synthesis and autocrine activation result in the dysregulation of growth factor receptors during oral carcinogenesis. Transforming growth factor alpha and beta (TGF-) expression is abnormal throughout the carcinogenesis process. TGF- is reportedly produced early in the development of OC, first in the hyperplastic epithelium that develops histologically, and then in the invasive carcinoma that develops in the inflammatory cell infiltrate, especially the eosinophils, that surrounds the infiltrating epithelium. TGF- has been discovered to be present in "normal" oral mucosa in individuals who later develop a second primary cancer. TGF- promotes cell proliferation by binding to EGFR and promotes angiogenesis.

Microscopically "normal" oral mucosa from individuals with head and neck cancer who subsequently develop second primary carcinomas overexpresses TGF-, indicating a "premalignant" lesion with fast epithelial proliferation and genetic instability. Patients with OCs who overexpress both TGF- and EGFR have been demonstrated to have a much worse prognosis than those who just overexpress EGFR (Nindl and Rösl 2009). The TGF receptors receive signal from GF-1 and translate it into action by phosphorylating SMAD2 and SMAD3. These phosphorylated proteins, when combined with SMAD4, control the transcription of the target genes.

Inhibitors of Apoptosis

Apoptosis, often known as "programmed cell death," is a physiological process in which cells die after a series of events once their purpose has been fulfilled. Any modifications to the process by which a cell undergoes apoptosis not only promote aberrant cell growth but also increase a cell's susceptibility to anticancer treatments like radiation and cytotoxic chemicals. The shift in expression of B-cell lymphoma-2 (Bcl-2) family members is one of the hypothesized pathways for the development of resistance to cytotoxic antineoplastic therapy.

The Bcl-2 family of proteins, which includes 25 pro- and anti-apoptotic members, maintains a balance between freshly developing cells and old dying cells. Apoptotic cell death can be stopped when the ratio of pro- to antiapoptotic proteins is altered or out of balance, leading to an overexpression of anti-apoptotic Bcl-2 family members. Cancer chemotherapy can be made more effective by focusing on the antiapoptotic Bcl-2 family of proteins. This will also help overcome drug resistance.

The intrinsic and extrinsic cell-death mechanisms are the two main apoptosis routes. The intrinsic cell death route, also known as the mitochondrial apoptotic pathway, is activated by a variety of signals, including radiation, cytotoxic medicines, cellular stress, DNA damage, and the removal of growth factors. It mostly causes apoptosis in response to internal stimuli. In this method, cytochrome c proteins are released from the mitochondrial membrane space, activating procaspase-9 and causing apoptosis (Kang and Reynolds 2009). The extrinsic cell-death mechanism carries out cascade activation of caspases and operates independently of mitochondria. Fas and tumor necrosis factor-related apoptosis-inducing ligand receptors are examples of cell-surface death receptors.

Role of Human Papillomavirus in Oral Cancer

HPV has a firmly recognized involvement in the etiology of human cancers. It is also thought to have a significant role in the development of mouth cancer. There are more than 100 different subtypes of HPV, some of which have been classified as high-risk HPVs because they have a role in oral carcinogenesis. Eighty five per cent of individuals have SCC. The malignant transformation is caused by the viral DNA integrating into the host genome. The virus has two oncogenes, E6 and E7, which can cause the overexpression of E6 and E7 proteins by interrupting the open reading frames of E1 and E2. The E2F version of the transcription factor is replaced by the degraded form of pRb, which then activates the gene that promotes cell growth. The beginning of HPV-mediated carcinogenesis is thought to occur when the E6 protein breaks down the p53 protein, which disrupts the cell cycle control in the infected cells. Since the virus is difficult to cultivate, evaluating its contribution to oral squamous cell carcinoma (OSCC) pathogenesis often involves employing PCR techniques to identify the viral DNA genome or the expression of viral genes. E6 and E7 have a significant role in the development of cervical cancer and the upper aerodigestive tract cancer that is caused by HPV.

Cellular signaling pathways are not isolated from one another but are linked together to form complex signaling networks. Any modification or diversity of this cellular signaling network, such as increased synthesis of growth factors or cell surface receptors, increased transcription, translation, or intracellular messenger levels, can result in aberrant cell proliferation and is one of the causes of multifactorial oral carcinogenesis. These alterations can then lead to proto-oncogene activation or loss of tumor suppressor action, resulting in a phenotype capable of enhancing cellular proliferation, reducing cell cohesion, and inducing local infiltration and metastasis (Figure 2.4).

ROLE OF BIOINFORMATICS IN ORAL CANCER

Early asymptomatic detection of OC improves cure rates and patient quality of life by reducing the need for extensive debilitating treatments. Unfortunately, at the time of diagnosis, more than half of patients

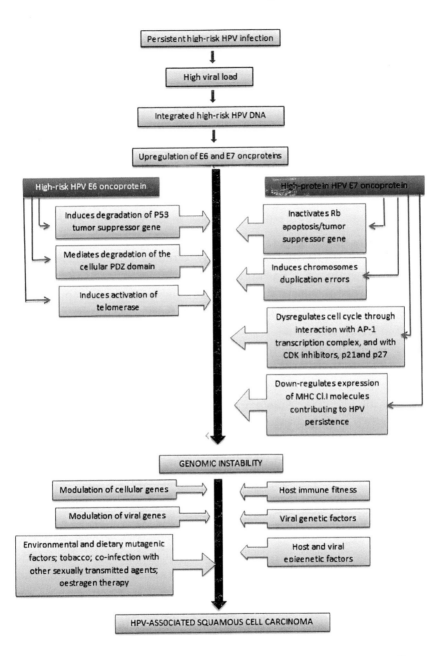

FIGURE 2.4   Flowchart of HPV pathogenesis in OSCC. *HPV*, Human papilloma-virus; *OSCC*, oral squamous cell carcinoma.

suffering from OC have evidence of spread to regional lymph nodes and metastases, and about two-thirds of patients have visible symptoms, which is a negative prognostic indicator. Although screening has been emphasized as a means of reducing the mortality and morbidity associated with OC, the difficulty in clinically distinguishing premalignant and malignant lesions from similar looking benign lesions makes visual identification of OC at premature stage difficult. Precancers and early stage OCs are difficult to detect by visual inspection alone, and even highly trained professionals with extensive experience may overlook and neglect them. As a result, a method of early, curable detection is critical, and it could lead to depletion in the currently unacceptable high rates of OC morbidity and mortality. Although automated cytology was proposed in the 1950s as a way to reduce false-negative results, early attempts that depended on algorithmic computers to analyze data were unsuccessful. New, nonalgorithmic neural network computers that were created in the late 1980s for missile defense were ultimately able to overcome this constraint.

One of the most important omics branches for experimental/clinical studies and applications is cancer bioinformatics. Precision medicine can benefit greatly from the availability of these data and the insights they may provide into disease biology. Cancer, according to Darwin's method, is caused by the gradual, inadvertent acquisition of alterations in the population of premalignant individual cells of genomes. Through the loss of general genome stability, these slow alterations gradually modify the phenotypic of normal cells, making them more cancerous.

Because of the pragmatic differences in their cell morphologies and unique prognoses, cancers of various organ and tissue systems are sometimes regarded as distinct illnesses. The discrepancies between two types of tumors, on the other hand, might be explained by underlying distinctions in their respective origin tissues. With the exception of constitutively expressing housekeeping genes that are shared across all tissues, each normal human tissue has its own distinct gene expression pattern. Further research indicated that housekeeping genes are not always expressed at the same level across all tissues; rather, each tissue appears to have its own housekeeping gene expression profile. In reality, housekeeping genes are less compact and developed than tissue-specific genes, and their coding and core promoter sequences change more slowly (Zhang and Li 2004). The differences in the functionality of the resultant proteins in a certain tissue reveal the distinct patterns of gene expression in various tissue types. Furthermore, alternative splicing of many human genes resulted in the production of multiple mRNA transcripts including various sets of exons. Alternative transcript expression patterns have been described in a wide range of human tissues, while their mechanism is unknown.

OSCC is a solid tumor that develops from aberrant squamous cells in the oral cavity. SCCs of the head and neck have the greatest rate of OSCC (80%). OSCC is characterized by epithelial tumor cell invasion into underlying tissues, and it is more common in the elderly. OSCC is generally asymptomatic until it has progressed to a late stage.

The diagnosis of OSCC is frequently delayed, and patients with advanced oral malignancies have a five-year survival rate of only 20%. OSCC is a major worldwide health problem that has yet to be resolved.

OSCC has a higher prevalence and a poor prognosis, which has prompted several investigations to look into the underlying mechanisms of OSCC development. p16, p15, hMLH1 (Human Mut-L Homologue 1), MGMT (O 6-methylguanine-DNA methyltransferase), and E-cadherin were shown to be abnormally methylated in the course of OSCC. These methylation genes were linked to a higher incidence of OSCC in patients. In addition, several cytokines and pathways have been linked to the development of OSCC. Interleukin-8 (IL-8) expression was found to be enhanced, and it is thought to have a role in oral squamous carcinoma cell invasion via modulating matrix metalloproteinase pump-1 (MMP-7) expression. Other cytokines and pathways linked to OSCC include cyclooxygenase-2, myeloid cell leukemia-1, the serine/threonine kinase signaling route, and the PD-1/PD-L1 (inhibitory receptor Programmed Death-1/Programmed cell death ligand pathway) (Vlatković and Boyd 2018).

The most effective way to improve survival rates is to detect OC early. The tumor, node, and metastasis classification and histopathological diagnosis are used to plan OC treatment. These methods are subjective, and they frequently lack the sensitivity needed to detect the disease early on. Furthermore, these methods do not account for tumor aggressiveness, prognosis, or treatment response.

As a result, biomarkers are urgently needed to identify high-risk individuals, improve cancer detection in the early stages, and predict disease outcomes and response to therapy. Rapid advances in high-throughput genomic and proteomic technologies have paved the way for a better understanding of the molecular pathogenesis of OC as well as the identification of candidate biomarkers.

CONCLUSION

The comparison, analysis, and interpretation of genetic and genomic data, as well as a broader understanding of the evolutionary elements of molecular biology, are made easier with the use of bioinformatics tools. It aids in the analysis and cataloging of biological networks and pathways This bioinformatics analysis presents a methodology that assists in the early detection and individualized treatment of OC. The study was able to identify five molecular subclasses of OC and the genes associated with each tumor stage.

REFERENCES

Brennan P, Lewis S, Hashibe M, Bell DA, Botteffa D, Bouchardy C, et al. Pooled analysis of alcohol dehydrogenase genotypes and head and neck cancer—review. Am J Epidemiol. 2004;159(1):1–16.

Denaro N, Lo Nigro C, Natoli G, Russi EG, Adamo V, Merlano MC. The role of p53 and MDM2 in head and neck cancer. ISRN Otolaryngol. 2011;2011:931813.

Ferlay JEM, Lam F, Colombet M, Mery L, Piñeros M, Znaor A, Soerjomataram I, Bray F. Estimating the global cancer incidence and mortality in 2018: GLOBOCAN sources and methods. Int. J. Cancer. 2019;**144**:1941–1953.

García-Martín JM, Varela-Centelles P, González M, Seoane-Romero JM, Seoane J, García-Pola MJ. Epidemiology of Oral Cancer. In: Panta P, editor. Oral Cancer Detect. 1st ed. Volume 3. Springer International Publishing; Cham, Switzerland: 2019. pp. 81–93.

Gudiseva S, Katappagari KK, Kantheti LPC, Poosarla C, Gontu SR, Baddam VRR. Molecular biology of head and neck cancer. Journal of Dr. NTR University of Health Sciences. 2017;6(1):1.

Jain A. Molecular Pathogenesis of Oral Squamous Cell Carcinoma. In: Daaboul HE, editor. Squamous Cell Carcinoma - Hallmark and Treatment Modalities. London: IntechOpen. 2019. Available from: https://www.inte-chopen.com/chapters/67447

Kang MH, Reynolds CP. Bcl-2 inhibitors: Targeting mitochondrial apoptotic pathways in cancer therapy. Clin Can Res. 2009;15(4):1126–1132.

Lyford-Pike S, Peng S, Young GD, et al. Evidence for a role of the PD-1:PD-L1 pathway in immune resistance of HPV-associated head and neck squamous cell carcinoma. Cancer Res. 2013;73(6):1733–1741.

Nindl I, & Rösl F (2009). Molecular pathogenesis of squamous cell carcinoma. Cancer treatment and research, 146:205–211. https://doi.org/10.1007/978-0-387-78574-5_18

Ram H, Sarkar J, Kumar H, Konwar R, Bhatt ML, Mohammad S. Oral cancer: risk factors and molecular pathogenesis. J Maxillofac Oral Surg. 2011;10(2):132–137.

Rivlin N, Brosh R, Oren M, Rotter V. Mutations in the p53 tumor suppressor gene: Important milestones at the various steps of tumorigenesis. Genes Cancer. 2011 Apr;2(4):466–474.

Shigeishi H, Ohta K, & Takechi M (2015). Risk factors for postoperative complications following oral surgery. Journal of applied oral science : revista FOB, 23(4):419–423. https://doi.org/10.1590/1678-775720150130

Sujir N, Ahmed J, Pai K, Denny C, Shenoy N. Challenges in early diagnosis of oral cancer: Cases series. Acta Stomatol Croatica. (2019) 53:174–180.

Watkinson J, & Clarke R (Eds.). (2018). Scott-Brown's Otorhinolaryngology and Head and Neck Surgery: Volume 3: Head and Neck Surgery, Plastic Surgery (8th ed.). CRC Press. https://doi.org/10.1201/9780203731000

Williams HK. Molecular pathogenesis of oral squamous carcinoma. Mol Pathol. 2000;53(4):165–172.

World Health Organization. Oral Health. [(accessed on 11 March 2020)];2018. Available online: https://gco.iarc.fr/today/data/factsheets/cancers/1-Lip-oral-cavity-fact-sheet.pdf

Wyss A, Hashibe M, Chuang SC, Lee YC, Zhang ZF, Yu GP, Winn DM, Wei Q, Talamini R, Szeszenia-Dabrowska N, et al. Cigarette, cigar, and pipe smoking and the risk of head and neck cancers: Pooled analysis in the international head and neck cancer epidemiology consortium. Am J Epidemiol. 2013;178:679–690.

Zhang LQ, Li WH. Mammalian housekeeping genes evolve more slowly than tissue-specific genes. Mol Biol Evol. 2004; 21:236–239.

CHAPTER 3

Biomarkers as Diagnostic Tool

Bhargav Shreevatsa[1], Viveka S[2], Sai Chakith M R[3], Chandan Dharmashekara[1], Anisha S Jain[4], Siddesh V Siddalingegowda[4], Umamaheswari S[4], Mahesh KP[2], Shiva Prasad Kollur[5], Chandan Shivamallu[1]

[1]*Department of Biotechnology and Bioinformatics, JSS Academy of Higher Education and Research, Mysuru, Karnataka, India*

[2]*Department of Oral Medicine and Radiology, JSS Dental College and Hospital, JSS Academy of Higher Education & Research, Mysuru, Karnataka, India*

[3]*Department of Pharmacology, JSS Medical College, JSS Academy of Higher Education and Research, Mysuru, Karnataka, India*

[4]*Department of Microbiology, JSS Academy of Higher Education and Research, Mysuru, Karnataka, India*

[5]*School of Physical Sciences, Amrita Vishwa Vidyapeetham, Mysuru, Karnataka, India*

INTRODUCTION

Biomarkers are biological molecules that equate with the presence or absence of a disease state, are prognostic, and indicate a tumor's response to a certain therapy. Global tumor evaluation approaches based on DNA and gene expression microarrays have provided new insights into altered molecules and pathways for future study in squamous cell carcinoma (SCC); nevertheless, their application as a biomarker for individual tumor assessment continues to be a challenge.

DOI: 10.1201/9781032625713-3

A tumor marker is a chemical that is present in or produced by a tumor, the tumor's host, or both in response to the presence of the tumor and that can be used to distinguish a tumor from normal tissue or to identify the presence of a tumor through blood or secretory tests (Chang et al. 2021). According to a different definition, tumor markers are "specific, novel or structurally altered cellular macromolecules or temporally, geographically, or quantitatively altered normal molecules that are associated with malignant neoplastic cells" (Lehto and Pontén 1989). Another team of researchers defined tumor markers as "cellular components that are improperly produced by malignancies and that may be detected in different body fluids and on the surface of cancer cells"(Casciato and Territo 2012).

Current revelations in molecular genetics have enhanced our understanding of the molecular mechanisms behind head and neck squamous cell carcinoma (HNSCC) development, leading to the identification and classification of different biomarkers. HNSCC biomarkers should help in the early detection of both primary and recurring tumors. A distinct pattern of gene expression is frequently reported during the progression from histologically normal tissue to primary carcinoma and nodal metastasis, with progressive up-or downregulation of expression, whereas alternatively expressed genes may be useful biomarkers for the prognostication of malignant transformation. Biomarkers are required to aid in early identification, risk assessment, and therapy response in HNSCC, which features a diverse variety of neoplasms. Biomarker discovery includes molecular, genomic, and phenotypic investigation of tumor samples and surrogate materials. Biomarkers should be very concise, objective, quantitative, and cost-effective, as well as precise and simple to use. Because of the heterogeneity of head and neck tumors, it is advisable to combine several selected markers with histological features for risk assessment.

In cancer patients, reverse transcription-polymerase chain reaction (RT-PCR)-based detection approaches have been used to identify circulating mRNA biomarkers in blood or plasma. Parallel to the large influx of such indications in human physiological fluids, more powerful and cost-effective methods for mass screening for genetic alterations are becoming more commonly available. According to a new study via microarray technology, a diverse panel of human mRNA exists in saliva, providing a distinct clinical method, salivary transcriptome, for disease diagnosis and normal health surveillance. Even though (epidermal growth factor receptor) EGFR is present in up to 90% of HNSCC patients, overexpression of EGFR has been related to a decreased overall survival and recurrence rate.

Because elevated EGFR was associated with a poor prognosis, it was one of the first biomarkers studied as a potential therapy for HNSCC. Cetuximab, a monoclonal antibody directed against the EGFR's extracellular receptor domain, inhibits ligand binding and downstream signaling while also contributing to the receptor's long-term downregulation. It is the most effective targeted therapy for HNSCC.

A viable biomarker has numerous advantages, which includes objective and quantitative evaluation, measurement precision, and dependability. Cancer biomarkers can also be used to screen for primary malignancies, evaluate disease risk, and differentiate between benign and malignant findings, as well as distinct forms of malignancy. Biomarkers can also be utilized to predict and monitor disease state and progression, as well as posttreatment disease reoccurrence and progression or therapeutic response (Santosh, Jones, and Harvey 2016). Aside from the benefits of biomarkers, potential issues and obstacles should be considered. Incorrect specimen collection, transportation, and storage resulted in laboratory measurement discrepancies.

Confounding variables that might interfere with biomarker testing should be recognized ahead of time. Internal variables include age, weight, gender, and metabolic parameters, while external influences can be discovered in detection kit batches. The cost and efficiency of a certain treatment, as well as the true influence on treatment results, must all be considered. Visual screening has already been demonstrated to be the most cost-effective technique of screening for oral cancer in high-risk groups. However, data and a comprehensive assessment of biomarker cost-effectiveness in clinical practice are lacking.

Oral Cancer Biomarkers

Any tumor marker's primary function is to identify the primary disease. The marker must be 100% specific and sensitive to obtain definitive results. A tumor marker that aids in diagnosis will be useful in determining the best treatment strategy. The majority of tumor markers are ambiguous and can be found in normal, benign, and cancerous tissue. They can, however, be used to make a differential diagnosis of suspicious lesions. Tumor markers are frequently used to determine the histogenetic origin of oral cavity neoplasms, allowing at least some of the entities on the differential diagnosis list to be ruled out. Tumor markers can be one-of-a-kind genes or gene products found only in tumor cells, or they can be gene products or genes found in normal cells that are aberrantly expressed in specific tumor cell locations. In response to cellular stress or environmental signals, they are present in abnormal amounts or function abnormally. Tumor markers might be distinctive genes or gene products present only in tumor cells, or they can be gene products or genes seen in normal cells but expressed abnormally in certain tumor cell sites. They are present at aberrant levels or operate improperly in response to cellular stress or external cues. Tumor indicators can be located within or on the surface of cells, or they can be discharged into the extracellular area, including the circulatory system. Tumor cells' unique biological properties, such as their capacity to infiltrate, spread, proliferate indefinitely and mitigate apoptosis, and angiogenesis, are regulated by complicated biochemical pathways, the components of which might be tumor indicators.

Chan and Sell (Shpitzer et al. 2009) have summarized the possible uses of tumor markers as follows:

- Screening in the general population
- Differential diagnosis in symptomatic patients
- Clinical staging of cancer
- Estimating tumor volume
- Prognostic indicator for disease progression
- Evaluating the success of the treatment
- Detecting recurrences
- Monitoring responses to therapy
- Radio immunolocalization of tumor masses
- Determining the direction for immunotherapy

Several salivary protein markers for oral squamous cell carcinoma (OSCC) have been assessed in several studies, with sensitivity and specificity values that are modest for prognostic prediction. Defensins are peptides that have cytotoxic and antibacterial properties. They are seen in the azurophil granules of polymorphonuclear leukocytes. Salivary defensin-1 levels were shown to be higher in patients with OSCC compared to healthy controls, which were thought to be symptomatic of the presence of OSCC. Interleukin-8 was detected in greater concentrations in saliva, whereas

IL-6 was found in higher concentrations in the serum of OSCC patients. As a consequence, they concluded that IL-8 in saliva and IL-6 in serum had the potential to be OSCC indicators. Other salivary biomarkers that have significantly altered in OSCC patients when compared to healthy controls include inhibitors of apoptosis (IAP), SCC-associated antigen (SCC-Ag), carcinoembryonic antigen (CEA), carcino-antigen (CA19-9), CA128, and serum tumor marker (CA125).

An ideal tumor marker should meet certain criteria:

- Be simple and cheap to measure in widely available bodily fluids.

- Be significant to the tumor under investigation and primarily related to it.

- A stoichiometric link exists between the marker's plasma level and the related tumor mass.

- Have abnormal urine, plasma, or both levels in the presence of micro-metastases, that is, when no clinical or currently available diagnostic tests disclose their presence.

- Have stable plasma, urine, or both levels that are not vulnerable to wild variations. If present in the plasma of healthy persons, the concentration is substantially lower than that seen in conjunction with all stages of cancer.

Classification of Oral Cancer Biomarkers

Biomarkers are categorized based on disease stage, biomolecules, or other parameters. Clinical and experimental research is increasingly focusing on OSCC screening and early detection, while clinical and experimental studies are focusing on diagnostic signs. Diagnostic indicators can be found at any stage of cancer development. Several attempts have been made using diverse approaches to find and categorize cancer biomarkers, but no broad agreement has been obtained.

Cancer biomarkers can be any biologically generated component or process that leads to a cancer diagnosis, during the stage of diagnosis (in the therapy and treatment module), or after diagnosis (in the therapy and treatment module). As a result of the great expansion of knowledge over the last several decades, several methods of categorizing cancer biomarkers have been offered in many disciplines of biomedical sciences and technology development.

A cancer diagnostic marker can be tuned to the patient's tissue, stage, follow-up, recurrence, and age. Despite efforts to identify cancer biomarkers, no consensus has been reached. Recent efforts, however, have focused on novel and noninvasive techniques utilizing human saliva collection, such as proteome, proteins, transcriptomic, and metabolomic markers, for the diagnosis and understanding of the OSCC genetic architecture.

Prediction, Detection, Diagnostic, and Prognostic Cancer Biomarkers

Prognostic biomarkers: Prognostic biomarkers are based on the characteristics that differentiate benign from malignant cancers. These biomarkers may also be formulated based on tumor differentiation status, which might impact physicians' treatment modalities selections. These indicators are very useful in predicting the recurrence of oral cancer. MammaPrint (Agendia), Oncotype DX (Genomic Health), and the H/I (AviaraDx) are prominent commercially produced assays for determining clinical outcomes following surgery based on genetic expression readout.

Predictive biomarkers: Predictive biomarkers, also known as response indicators, are primarily used to assess the impact of dispensing a certain treatment. These factors allow doctors to select the appropriate mix of chemotherapeutic medications for a specific patient.

Detection biomarkers: Cancer markers used to choose chemotherapeutic drug doses in a given set of tumor-patient settings are known as detection indicators. These factors assist in decreasing cancer therapeutic doses below the level of toxicity and driving clinical trials forward.

Diagnostic biomarkers: Diagnostic indicators can appear at any point throughout the progression of cancer. Furthermore, a diagnostic cancer marker may be tailored to a patient's stage, tissue, recurrence, follow-up, and age (Casiato & Lowitz 1983).

Cancer Biomarkers Based on Biomolecules
DNA Mutations in DNA nucleotides are tumor promoters (APC, Ras), cell cycles (cyclins), tumor suppressors (p16, p19, p53, Rb), and DNA-repair associated genes (XRCC), which are connected to cancer prognosis and diagnosis, but their clinical relevance remains unknown. DNA can be found in tissue, saliva, serum, sputum, bronchial tears, tumor cells, and cerebrospinal fluid circulating in the blood and bone marrow (Mishra and Verma 2010). Alterations in mtDNA (mitochondrial DNA)

molecules, in addition to nuclear mutations, have been widely advocated as potential biomarkers for a variety of malignancies. Carcinogenesis depends on the epigenetic alteration of nucleic acids and related proteins (non-histones and histones). DNA mismatch-repair genes are controlled by histone deacetylation, promoter region CpG methylation, and lysine-specific histone-H3 methylation; tumor-suppressor genes (CDKN2A, T P53, APC, and BRCA1) are the genes most affected by these changes (MLH1 or the O6-methyl-guanine-DNA methyltransferase gene, MGMT). CpG methylation-mediated gene silence is one of the most well-studied epigenetic alterations to date. The severity of the lesions is strongly linked to the degree of methylation in prostate cancer tissue, sputum/serum from individuals with lung cancer, and saliva from those with oral malignancies.

RNA and MicroRNA (miRNA)   Quantitative RT-qPCR, serial analysis of gene expression, differential display, bead-based approaches, microfluid card and micro-array analysis are a few techniques used to find cancer biomarkers at the RNA expression level. Laser capture-based microscopy is implemented in numerous grades and stages of therapy to seek pure RNA signatures. Heat maps supervised algorithms and snapshots are utilized to compare RNA expression, which is then linked to diagnosis and prognosis. miRNAs (MicroRNAs) are small noncoding RNAs. For various cancer forms, the expression of distinct miRNA populations is linked to clinical features in a tissue- and time-dependent way. There is enough data to show that miRNA expression patterns may be used to categorize human malignancies, implying that illness prediction and therapy outcome are linked.

In the realm of oncogenesis, the field of metastasis-associated miRNA markers is rapidly expanding, and these markers have recently been coined "metastasis." MiRNA has the potential to act as both a tumor suppressor and an oncogene. MiRNAs are employed as a biomarker in cancer patients for diagnosis, prognosis, staging, clinical risk and prediction, and clinical intervention.

Protein Markers   Protein-based biomarkers are more relevant than DNA- or RNA-based biomarkers since proteins are the primary governing biomolecules in cells. Proteomic indications are closer and more relevant to illness initiation and development because protein molecules impact molecular pathways in altered and normal cells. The only

FDA-approved biomarkers that are presently accessible for therapeutic usage appear to be protein molecules. Protein-based signatures are produced using high throughput platforms like matrix-associated laser absorption desorption ionization time of flight (MALDI-TOF), mass spectroscopy (MS), surface-enhanced laser absorption desorption ionization time of flight, and reverse phase microarray, as well as traditional two-dimensional fluorescence difference gel electrophoresis, polyacrylamide gel electrophoresis, and new technologies like quantum dots and nanoparticles have been created to examine the viability of using protein molecules as cancer biomarkers.

Carbohydrate Biomarkers   The expression of some N-linked and O-linked glycans alters as cancer progresses in some cases. These changed glycoforms might be used as potential cancer biomarkers. The most common method for detecting disease-related carbohydrate markers is MS. Serum glycomics has recently been used to detect esophageal cancer. Because glycomarkers (proteoglycans glycolipids and glycoproteins) are more stable than RNA and proteins, they are better suited to epidemiological studies in which human populations are screened to determine who is most likely to develop cancer throughout their lives. Profiling O- and N-linked glycosylation of protein molecules at serine and threonine residues in human sera, tissue, and cancer lines using MALDI-TOF and electrospray ionization is an essential method for detecting glycan-based cancer biomarkers. Modulated expression of glycosyltransferases causes increased branching and changed terminal structures of glycans (sialyl and fucosyl-transferases).

Pathogenic Cancer Biomarkers
Viral Biomarkers   Infectious agents, especially viral infection, cause 15–20% of all human cancers. Therefore, the association of certain tumor types with viruses makes these pathogens particularly intriguing biomarkers. Infection with the Epstein–Barr virus (EBV) has been linked to nasopharyngeal carcinoma (NPC) and lymphoma. Human papillomavirus (HPV) is connected to cervical cancer and subgroups of head and neck malignancies. Other viruses connected to cancer include Kaposi's sarcoma-associated herpesvirus (KSHV) and human herpesvirus 8 (HHV-8) for sarcoma and lymphoma, respectively, and RNA viruses including human T-cell lymphotropic virus type 1 (HTLV-1) for specific forms of leukemia.

Bacterial Biomarkers *Helicobacter pylori* (*H. pylori*) causes persistent low-level stomach lining inflammation. Infection with *H. pylori* is closely correlated with the presence of duodenal and gastric ulcers, and it is a well-established biomarker for gastric cancer. *H. pylori* is detected in patients using either DNA polymorphisms or antibody-based methods.

Imaging Biomarkers For the early detection of cancer, physical tests and noninvasive technology are not always sufficient. Modern imaging methods, including X-rays, computed tomography, ultrasound, radionuclide imaging, and magnetic resonance imaging, are frequently utilized for cancer screening, diagnosis, and treatment effectiveness evaluation.

Mesenchymal Markers

- Muscle antigens—Desmin, Actin, Myoglobin, Myosin.
- Vasculator antigens—CD34, CD31.
  - Neural antigens—S100, NSE, GFAP, Synaptophysin, Nerve growth factor receptors.

Prognostic Markers

- Cell adhesion molecules—Cadherins, Integrins, Selectins.
- Proliferative markers—Ki67, PCNA, AgNORs.
- Biochemical markers
- Enzymes and isoenzymes—PAP, PSA, PALP, Lysozyme.
- Protein—Ferritin, Beta-protein, Glycoprotein, Immunoglobulins.
- Hormone receptors—Progesterone receptor, Estrogen receptor
- Epithelial markers.

Malati's Classification is as Follows

- Oncofetal antigens (e.g., alpha-fetoprotein [AFP], carcinoembryonic antigen [CEA], pancreatic oncofetal antigen, and fetal sulfoglycoprotein)
- Tumor-associated antigens/cancer antigens, for example, CA125, CA19-9, CA15-3, CA72-4, and CA50

- Hormones, for example, beta human chorionic gonadotropin, calcitonin, and placental lactogen

- Hormone receptors (e.g., estrogen and progesterone receptors)

- Enzymes and isoenzymes (e.g., prostate-specific antigen [PSA], prostatic acid phosphatase [PAP], neuron-specific enolase [NSE], glycosyl transferases, placental alkaline phosphatase [PALP], terminal deoxynucleotidyl transferase, lysozyme, alpha-amylase

- Serum and tissue proteins (beta-2 microglobulin, monoclonal immunoglobulin/para proteins, glial fibrillary acidic protein [GFAP], protein S-100, ferritin, and fibrinogen degradation products)

- Other biomolecules, for example, polyamines (Figure 3.1)

Reddy et al.'s Classification Is as Follows:

- Proliferative markers: Ki-67, proliferating cell nuclear antigen (PCNA), p27 Kip/gene, DNA polymerase alpha, p105, p120, and stain

- Oncogenes: c-erb-gene, ras gene, myc gene, and bd-2 gene

- Growth factors and receptors: EGFR, transforming growth factor-β, hepatocellular carcinoma, fibroblast growth factor (FGF) receptor, insulin, and insulin-like growth factor receptor

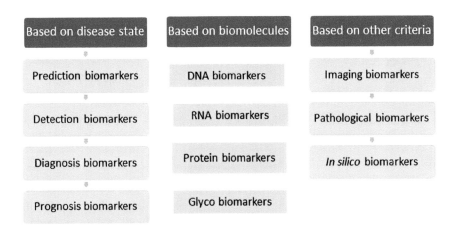

FIGURE 3.1   Classification of cancer biomarkers.

- Tumor suppressor genes: p53, retinoblastoma susceptibility suppressor gene

- Serological tumor markers: Markers associated with cell proliferation, markers related to cell differentiation (carcinoembryonic proteins such as carcinoembryonic Ag, α-fetoprotein), markers related to metastasis, markers related to other tumor-associated events, markers related to malignant transformation, inherited mutation, monoclonal Ab-defined tumor markers

Schliephake's Classification Is as Follows:
Tumor Growth Markers

- Epithelial growth factor (EGF)

- Cyclin

- Nuclear cell proliferation antigens

- Argyrophilic nucleolar organizer region (AgNORs)

- S-phase kinase-interacting protein 2

- HSP 27 and 70 (heat shock protein)

- Telomerase

Markers of Tumor Suppression and Antitumor Response

- Retinoblastoma protein (pRb)

- Cyclin-dependent kinase inhibitors

- p53

- bax

- Fas/FasL

Angiogenesis Markers

- Vascular endothelial growth factor/receptor

- Platelet-derived endothelial cell growth factor

- FGFs

Markers of Tumor Invasion and Metastatic Potential

- Matrix metalloproteases
- Cathepsins
- Cadherins and catenins
- Desmoplakin

Cell Surface Markers

- Carbohydrates
- Histocompatibility antigen
- CD57 antigen

Intracellular Markers

- Cytokeratins.

Markers of Anomalous Keratinization

- Filagrins
- Invoulcrin
- Desmosomal proteins
- Intercellular substance antigen
- Nuclear analysis

Arachidonic Acid Products

- Prostaglandin E2
- Hydroxyeicosatetraenoic acid
- Leukotriene B4
- Enzymes
- Glutathione S-transferase

Manikantan et al.'s Classification Is as Follows:

- Epithelial markers

- Cytokeratins

- Epithelial membrane antigen

- Oncofetal antigens

- AFP

- CEA

- Desmoplakin

Advantages of Using Tumor Markers
Cancer Screening and Early Detection
While early detection refers to finding cancer when it is still in an early stage, screening involves checking for cancer in people who do not exhibit any symptoms. The majority of tumor markers have not been shown to identify cancer much sooner than they would have otherwise been discovered, despite the fact that they were initially developed to screen for cancer in people who did not have symptoms. As a result, relatively few tumor indicators have been proven to be helpful in this regard.

Aid in Cancer Diagnosis
In most instances, cancer can only be detected by a biopsy, and tumor markers are rarely employed to do so. Tumor markers, on the other hand, can assist certain patients to assess the likelihood that they have cancer. It can also aid in the diagnosis of the etiology of cancer in individuals with advanced extensive illness.

Monitor Therapeutic Response
One of the most significant applications for tumor markers is in the monitoring of cancer patients. If the originally elevated tumor marker level decreases with therapy, it shows that the treatment is effective and has a positive effect. If, on the other hand, the marker level rises, the treatment is most likely ineffective and a change in treatment should be explored.

Prognostic Indication of Disease Progression
Some newer tumor indicators can help predict how aggressive cancer will be or how well it will react to certain therapies.

Indicate Any Relapses throughout the Follow-Up Period
During the follow-up period, indicate relapse. Cancers that relapse after the first therapy are also detected using markers. Some tumor markers may be useful after therapy is complete and there is no sign of residual malignancy. Prostate-specific antigen (for prostate cancer), human chorionic gonadotropin (for gestational trophoblastic tumors and germ cell malignancies of the ovaries and testicles), and cancer 125 are just a few examples (for epithelial ovarian cancer).

There are three levels of molecular markers for OSCC diagnosis: Changes in cellular DNA cause (I) changed mRNA transcripts, which cause (II) altered protein levels.

Changes in the Cellular DNA
Point mutations, translocations, deletions, amplifications, cyclin D1, methylations, microsatellite instability, EGFR, and the presence of HPV are some of the modifications that can occur in the host DNA of cancer or dysplastic cells. OSCC has been connected to chromosome 9p allelic deletion. Mitochondrial DNA alterations have also been shown to identify exfoliated OSCC cells in saliva. They have been spotted in 46% of head and neck malignancies. Direct sequencing alone found identical mitochondrial DNA alterations in 67% of OSCC patients' saliva samples. P53 gene alterations are found in almost half of all head and neck malignancies.

Altered mRNA Transcripts
Point mutations, deletions, translocations, amplifications, methylations, cyclin D1, EGFR, microsatellite instability, and the presence of HPV are among the host DNA changes that dysplastic or cancer cells experience. OSCC has been connected to chromosome 9p allelic deletion. Mitochondrial DNA alterations have also been shown to identify shed OSCC cells in saliva. They have been recognized in 46% of head and neck malignancies. Direct sequencing alone found identical mitochondrial DNA alterations in 67% of OSCC patients' saliva samples. P53 gene alterations are found in almost half of all head and neck malignancies. There are seven mRNA molecules in total transcripts of:

2.1. Interleukin 8 (IL-8), among other things, is involved in angiogenesis, replication, the calcium-mediated signaling system, cell adhesion, cell cycle arrest, chemotaxis, and immunological response.

IL-8 (also known as a neutrophil chemotactic factor) is a proinflammatory cytokine. This gene spans 3211 bp and is found on chromosome 4. Following translation, an mRNA of 99 amino acids and a weight of 11 kd is created. It is essential for tumor angiogenesis, cell cycle arrest, immunity, and cell adhesion. Electrochemical sensors were employed for detection, and Receiver-Operating Characteristic (ROC) analysis was performed to estimate predictive power. This test has demonstrated good specificity and sensitivity for IL-8 and IL-8 mRNA, both alone and in combination.

2.2. Interleukin-1B (IL1B), a signaling molecule implicated in proliferation, inflammation, and apoptosis. Interleukin-1 beta is a cytokine of the Interleukin 1 family. This protein has a role in inflammation, differentiation, apoptosis, and cell proliferation. The gene that codes for this protein is found on chromosome 2. Serum levels of IL IB are increased in patients with OSCC.

2.3. DUSP1 (dual specificity phosphatase 1) is a protein modification enzyme that aids in signal transmission and the response to oxidative stress. The DUSP1 gene on chromosome 5 encodes the enzyme dual specificity protein phosphatase 1. This gene has four exons separated by three introns and spans approximately 3111 bp. This gene's transcription produces an mRNA, which plays a key role in activating the MAPK pathway, which is involved in oxidative stress, protein modification, and signal transmission. DUSP1 is also regulated by the p53 gene, and hypermethylation of the DUSP1 gene is required for oral carcinogenesis.

2.4. H3F3A's DNA-binding activity (H3 histone, family 3A). The H3 histone, Family 3A protein is encoded by the H3F3A gene on chromosome-1. Histones are nuclear proteins that are important for the structural integrity of the chromosomal nucleosome. The H3F3A gene is made up of five exons that span around 9282 kb. The H3F3A mRNA is a proliferative marker with 135 amino acids and a molecular weight of 15 kd.

2.5. OAZ1 participates in polyamine biosynthesis (ornithine decarboxylase antizyme 1).

2.6. S100 calcium-binding protein P (S100P) is a protein and calcium ion binding protein.

2.7. SAT (spermidine/spermine N1-acetyltransferase) is an enzyme and a transferase.

Altered Protein Markers

Several salivary protein indicators for OSCC have been examined in a variety of studies with moderate sensitivity and specificity values for prognosis prediction. Given example, defensins are peptides with antibacterial and cytotoxic effects. They are present in the azurophil granules of polymorphonuclear leukocytes. Salivary defensin-1 levels were shown to be greater in OSCC patients compared to healthy controls, suggesting that higher levels of salivary defensin-1 indicate the existence of OSCC. SCC-Ag, IAP, CEA, carcino-antigen (CA19-9), serum tumor marker (CA125), CA128, intermediate filament protein (Cyfra 21-1), reactive nitrogen species, tissue polypeptide specific antigen and 8-OHG DNA damage marker, lactate dehydrogenase.

Metabolic Biomarkers

OPLS-DA model with Splot was used to select a panel of five salivary metabolites comprising -aminobutyric acid, n-eicosanoic acid, valine lactic acid, and phenylalanine. The combination of lactic acid, valine, and phenylalanine produced acceptable accuracy (0.89%, 0.97%), sensitivity (86.5% and 94.6%), specificity (82.4% and 84.4%), and positive predictive value (81.6% and 87.5%) in differentiating OSCC. Elevated amounts of hypoxanthine, guanosine, guanine, trimethylamine N-oxide, pipecolate, spermidine, and methionine have all been observed in saliva and have been used to screen controls from OSCC patients.

Eph and/or ephrin expression is found in a range of primary malignancies, including OSCCs. Ephrin and Eph receptors exhibit a range of biological actions in cancer, but their involvement in EFNB2/EphB4 signaling is thought to be associated with differentiation, angiogenesis, and development. As a consequence, EFNB2 gene expression may be a useful biological marker in predicting prognosis in patients with OSCC. According to Shpitzer et al., the levels of phosphorylated-Src, 8-oxoguanine DNA glycosylase and mammary serine protease inhibitor (Maspin) in the saliva of Oral-SCC patients decreased. Interleukin (IL)-6 and IL-8, which are well-known as post-inflammatory cytokines, have been found to dramatically increase in patients with OSCC, implying that they could be used as a diagnostic marker for premalignant lesions and oral malignant. According to Arellano-Garcia et al., IL-8 and IL-1 levels in OSCC patients were considerably greater. Several gene promoters are hypermethylated in head and neck cancer (HNC). Rosas et al. revealed that at least one of three genes in OSCC have aberrant methylation (p16, MGMT, or DAP-K). In the matching saliva sample of 65% of OSCC patients, aberrant promoter hypermethylation was also identified. Cyclin D1 gene amplification has been linked with poor outcomes in OSCC. Ki67 markers were found to be higher in the saliva of OSCC patients, but phosphorylated-Src, 8-oxoguanine DNA glycosylase and mammary serine protease inhibitor (Maspin) indicators were decreased.

A combination of salivary indicators increased the detection of oral cancer. A salivary mixture of mRNAs showed a discriminatory capacity of 91% sensitivity and specificity for oral cancer screening. Using saliva samples, methylation-specific PCR has proven to be a highly specific method for HNSCC identification (96%). Salivary transcriptomes such as IL-8, IL-IB, DUSP1, HA3, OAZ1, S100P, and SAT were reported to be highly specific (91%) and sensitive (91%). Salivary actin and myosin detection of oral cancer was 100% specific and 75% sensitive. A combination of salivary IL-1B, IL-8, OAZ, and SAT mRNA biomarkers can also accurately predict OSCC from precancerous individuals. As a consequence, a combination approach of salivary biomarkers might be employed as a screening tool to increase the early identification and diagnostic precision of oral precancer and malignancy.

DNA Methylation Biomarker

- ZNF582 and PAX1

ZNF582 belongs to the Krüppel associated box zinc finger proteins transcriptional regulator family. Recent studies have found that ZNF582 suppresses the growth of NPC via controlling the transcription and expression of the adhesion molecules NRXN3 and Nectin-3. Hypermethylation of ZNF582 enhances NPC metastasis by regulating these adhesion molecules. ZNF582 has also been identified to have a tumor suppressor role in anal cancer and esophageal carcinoma, and its methylation may be exploited as a cancer detection biomarker. PAX1 is renowned for its paired-box domain, and it works as a tumor suppressor gene (TSG) as well as a developmental gene during embryogenesis. Oral exfoliation cells from a suspect lesion are collected, and genomic DNA is extracted and converted by DNA bisulfite, which converts unmethylated cytosine to uracil while leaving methylated cytosine unaltered. After the DNA bisulfite conversion is finished, methylation-specific PCR is carried out to find out how much of each gene's cytosines are methylated. ZNF582 and PAX1 were first chosen from a multigene panel of SCC-type malignancies that contained ZNF582, PAX1, SOX1, NKX6.1, and PTPRR genes; they were then evolved into a technique for detecting oral dysplasia and oral cancer using oral scrapings (Zhao et al. 2020).

mRNA Biomarker

• Multipanel mRNA OAZ1, SAT, and DUSP1

Ornithine decarboxylase antizyme, OAZ1, is activated by a polyamine-dependent mechanism and has been associated with human OSCC cell line metastatic potential. It has been linked to the metastatic potential of human OSCC cell lines and is vital in DNA repair. SAT is the rate-limiting acetyltransferase in the catabolic route of polyamine metabolism. By catalyzing the acetylation of spermidine and spermine, the enzyme controls the intracellular concentration of polyamines and their transit out of cells. SAT expression has been reported to be considerably greater in prostate cancer and oral cancer. DUSP1 is a cysteine-based protein tyrosine phosphatase subtype that participates in several signaling pathways. Salivary DUSP1 mRNA levels in OSCC patients are substantially greater than in healthy controls. The three cell-free salivary mRNAs were coupled with housekeeping genes MT-ATP6 and RPL30 to form the CLIA-approved SaliMarkTM OSCC test (PeriRx, Pennsylvania, PA, USA). The SaliMarkTM OSCC test is a risk stratification test that should be used by clinicians when worrisome lesions are discovered. Saliva is collected with the SaliMarkTM saliva collection kit, and RNA is extracted and analyzed with RT-qPCR.

MicroRNA Biomarker

miR-125a—The gene that encodes miR-125a has 86 bases and is located on chromosome 19. miR-125a regulates cell proliferation and can influence genes involved in MAPK metabolism.

miR-200a—The gene that encodes microRNA-200a is found on chromosome 1, and there is proof that it has a role in early metastasis and tumor suppression. The miR-200a belongs to the miR-200 family of short RNA molecules, which also includes the miR-200a, miR-200b, miR-200c, miR-141, and miR-429. Its levels decrease during metastasis and are negatively associated with the extent of invasion.

miR-31—miR-31 is encoded by a 71-base gene that is found on the chromosomes. This is a tumor suppressor microRNA that is entirely depleted in metastatic oral cancers.

Protein Biomarker

- CD44

CD44 is a glycoprotein found on the cell's surface that affects several cellular functions, including cell proliferation and migration. It is also a tumor initiating and stem cell-associated biomarker linked to tumor start in a range of cancers, including oral cancer. CD44 is found largely in the basal and suprabasal areas of normal epithelial cells. As dysplasia proceeds, CD44 expression drifts into the superficial layers, implicating it in the early phases of carcinogenesis. Furthermore, CD44 is more numerous on the surface of relapsed tumors than on the surface of original tumors. Metalloproteinases that are overexpressed in advanced oral malignancies break CD44 and release it from the cell surface in a soluble form (solCD44).

A point-of-care (POC) IVD to detect the presence of CD44 in a saliva sample was developed to assess the risk of oral cancer. The BeVigilantTM Rapid Test (Florida, FL, USA Vigilant Biosciences) is a CE-approved POC IVD that identifies soluble CD44 (sCD44) and total protein levels in patients with oral cancer (Sun et al. 2020).

This POC is a dual-marker lateral flow test with two strips attached with monoclonal antibodies to CD44 and total protein. One monoclonal antibody recognizes sCD44, while the other recognizes a distinct binding region of human CD44. Using a fast reader and the BeVigilantTM Portal, physicians obtain qualitative data showing a low, moderate, or increased amount of the biomarker. A saliva sample is collected in a collection cup with 5 mL saline; the POC test strips are then put into the saliva sample until moist, and then set flat on an absorbent pad. A visible line will appear when the concentration of CD44 reaches a predetermined level.

- S100A7

S100A7 is a calcium-binding protein that belongs to the multigenic calcium-modulated S100 family and is found in the upper well-differed spinous layer of normal epithelium. Overexpression of S100A7 has been associated with the development of cancer in high-risk dysplastic oral lesions.

StraticyteTM (Proteocyte AI, Toronto, Canada) is an S100A7 test created in a laboratory. This test quantifies the S100A7 biomarker in biopsy tissues of patients at risk of oral cancer using StraticyteTM proprietary algorithms, sophisticated imaging, and digital pathology. The examination of histopathology begins with the collection of a biopsy specimen. Simultaneously, the number of S100A7 contained in the tissue may be quantified. S100A7, prothymosin alpha, 14-3-3, 14-3-3, and heterogeneous nuclear ribonucleoprotein K were discovered and validated using proteomic techniques to identify oral dysplasia and oral malignancies from normal oral tissues. S100A7 cytoplasmic overexpression stood out among other protein biomarkers as the most important potential marker associated with cancer progression in dysplastic lesions (Chan & Sell 1994).

Over five years, the single protein biomarker S100A7, StraticyteTM, was able to predict the chance of dysplastic lesions developing into cancer (Franzmann and Donovan 2018). When compared to histological dysplasia grading, StraticyteTM demonstrated improved objectivity, sensitivity, and predictive power. StraticyteTM demonstrated a 95% sensitivity and a 78% negative predictive value (NPV) (low risk vs. intermediate and high risk, respectively), whereas histopathological dysplasia had a sensitivity of 75% and an NPV of 59%. By objectively measuring S100A7, StraticyteTM was able to better characterize the risk of developing OSCC than histopathological dysplasia grade alone (Table 3.1).

Tumor Suppressor Genes

Oral cancer is caused by several reasons other than oncogenes. TSGs are cellular negative regulators that are inactivated during the transition of a premalignant cell into a malignant cell. TSGs are hypothesized to have a role in the genesis of cancer and are commonly inactivated by point mutations, rearrangements, and deletions in both gene copies. The TSG p53 gene is mutated in around 70% of adult solid tumors. The p53 protein suppresses cell division during the G1-S phase, increases DNA repair following DNA damage, and initiates apoptosis. These tasks are enabled by the p53 protein's capacity to influence the expression of various genes, including the WAF1/CIP gene, which produces the p21

TABLE 3.1   Molecular Markers for Diagnosis of OSCC

Changes in the Cellular DNA	Altered mRNA Transcripts	Altered Protein Markers
Allelic loss on chromosomes 9p	Presence of IL-8	Elevated levels of defensin-1
Mitochondrial DNA mutations	Presence of IL1B	Elevated CD44
p53 gene mutations	DUSPI (dual specificity phosphate 1)	Elevated IL-6 and IL-8
Promoter hypermethylation of gene (p16, MGMT, or DAP-K)	H3F3A (H3 histone, family 3A)	Inhibitors of apoptosis (IAP)
Cyclin D1 gene amplification	OAZI (ornithine decarboxylase antizyme 1)	Squamous cell carcinoma-associated anti-gen (SCC-Ag)
Increase of Ki67 markers	S100P (S100 calcium binding protein P)	Carcino-embryonic antigen (CEA)
Microsatellite alterations of DNA	SAT (spermidine/spermine N1-acetyltransferase)	Carcino-antigen (CA19-9)
Presence of human papillomavirus		CA128
		Serum tumor marker (CA125)
		Intermediate filament protein (Cyfra 21-1)
		Tissue polypeptide specific antigen (TPS)
		Reactive nitrogen species (RNS)
		8-OHdG DNA damage marker
		Lactate dehydrogenase (LDH)
		Immunoglobulin (IgG)
		s-IgA
		Insulin growth factor (IGF)
		Metalloproteinase MMP-2 and MMP-11

protein. p21 inhibits cyclin and cyclin-dependent kinase complexes. Tobacco use and smoking have been associated with p53 gene mutations in HNC. Another TSG, the doc-1 gene, is altered in malignant oral keratinocytes, resulting in decreased gene expression and protein activity (Tables 3.2–3.7).

TABLE 3.2 Transcriptomic Biomarkers Identified in Unstimulated Whole Saliva for OSCC Detection

Candidate Biomarkers	Techniques	Clinical Significance
IL-8, IL-1b	ELISA	Cell adhesion, immune response, signal transduction, inflammation, angiogenesis, apostasies, chemotaxis, replication and propagation
(DUSP1) Dual specificity phosphate 1	(qPCR) Quantitative PCR and microarrays followed by qPCR	Signal transduction, protein modification, oxidative stress
(H3F3A) H3 histone family 3A	(qPCR) Quantitative PCR and microarrays followed by qPCR	DNA-binding activity
Long non-coding HOTAIR (HOX transcript antisense RNA)	(qPCR) Quantitative PCR and microarrays followed by qPCR	Expression of HOTAIR is associated with p53 gene and causes DNA damage
miR-31, miR-125a, miR-200a	(qPCR) Quantitative PCR and microarrays followed by qPCR	Cellular growth, posttranscriptional regulation by RNA silencing complex and proliferation in elevated levels in OSCC

Miscellaneous Salivary Biomarkers for Oral Squamous Cell Carcinoma Detection

Oxidative stress-related chemicals, glucocorticoids, glycosylation-related molecules, and inorganic substances have all been linked to OSCC detection. HPLC and commercially accessible colorimetric assays

TABLE 3.3 Metabolomics Biomarkers Identified in Unstimulated Whole Saliva for OSCC Detection

Candidate Biomarkers	Techniques	Clinical Importance
Cadaverine, alanine, serine, glutamine, piperidine, taurine piperidine, choline, pyrroline hydroxycarboxylic acid, beta-alanine, alpha-aminobutyric acid betaine, tyrosine, leucine þ isoleucine, histidine, tryptophan, glutamic acid, threonine, carnitine, pipercolic acid, lactic acid, phenylalanine and valine	Capillary electrophoresis time-of-flight mass spectrometry (CE-TOF-MS). and HPLC with quadrupole/TOF MS	Improves diagnosis and prognosis of OSCC by assisting clinical detection
Hypoxanthine, trimethylamine N-oxide, guanine, guanosine, spermidine, methionine pipecolate.	CE-TOF-MS	Used as noninvasive oral cancer screening

TABLE 3.4 Protein Biomarkers Identified in Unstimulated Whole Saliva for the Detection of OSCC

Candidate Biomarkers	Techniques	Clinical Significance
Interleukin-6 (IL-6), interleukin-8 (IL-8), interleukin 1a (IL-1a), interleukin 1b (IL-1b), TNF-a, tissue polypeptide antigen (TPA), Cyfra 21-1, cancer antigen 125 (CA 125), telomerase, Mac-2 binding protein (M2BP)	ELISA (Enzyme-linked immunosorbent assay)	These pro-inflammatory and proangiogenic cytokines have been outlined as indicators of carcinogenic transformation from Oral precancerous lesions to Oral cancer. Telomerase activity is seen in tumor cells and is responsible for maintaining telomere length throughout chromosome replication. Cyfra 21-1, CA 125, and TPA markers are used as diagnostic tools. M2BP facilitates in the detection of OSCC.
CD44, CD59, Profilin, MRP14	Immunoblot	CD44 and CD59 have very high sensitivity and specificity in delineating cancer from benign diseases, whereas MRP14 is a calcium-binding protein
Glutathione	HPLC (High performance liquid chromatography)	The risk of developing OSCC is identified using an epidemiological marker for chemoprevention.
Mac-2 binding protein (M2BP), squamous cell carcinoma antigen 2, involucrin, calcyclin, cathepsin-G, azurocidin, transaldolase, carbonic anhydrase I, calgizzarin, myeloblastin, vitamin D-binding protein	ELISA, shotgun proteomics	M2BP may be used as a screening tool for noninvasive OSCC diagnosis.
Immunoglobulin heavy chain constant region gamma (IgG), S100 calcium binding protein, cofilin-1, transferrin, fibrin, a-1-antitrypsin (AAT)	LC/MS (Liquid chromatography/mass spectrometer) 2DE (2D electrophoresis)	Transferrin levels in saliva are related to tumor growth and stage. While fibrin is engaged in various carcinogenic pathways in OSCC. AAT is useful for the prediction and staging of OSCC.
Secretory leukocyte peptidase inhibitor (SLPI), cystatin A, keratin 36, thioredoxin, haptoglobin (HAP), Salivary zinc finger, Protein 510 peptide, a-amylase and albumin	MS-based proteomics	SLPI, cystatin A, and keratin 36 may be implicated in the prevention of OSCC. Thioredoxin mRNA levels are increased in oral carcinomas. Salivary zinc finger, protein 510 peptide, a-amylase, and albumin are all beneficial in detecting OSCC early.

TABLE 3.5 Genomic Biomarkers Identified in Unstimulated Whole Saliva for the Detection of OSCC (Santosh et al. 2016a,b)

Candidate Biomarkers	Techniques	Clinical Significance
Histone family 3 (HA3)	PCR, qPCR, and microarrays followed by qPCR	DNA-binding activity
S100 calcium-binding protein P (S100P)	S100 calcium-binding protein P (S100P)	Protein and calcium ion binding
spermidine/spermine N1- acetyltransferase EST (SAT)	PCR, qPCR, and microarrays followed by qPCR	Enzyme and transferase activity
ornithin decarboxylase antizyme 1	PCR, qPCR, and microarrays followed by qPCR	Polyamine biosynthesis
P53 gene codon	PCR, qPCR, and microarrays followed by qPCR	Detection of this gene at codon 63 gives fast, accurate, and sensitive diagnosis of OSCC
Loss of heterozygosity in combination of other biomarkers D3S1234, D17S79, and D9S156	PCR, qPCR, and microarrays followed by qPCR	An early indicator of the precancerous lesion that is most likely to be changed in malignancy
Mitochondrial DNAs such as cytochrome cooxidase I and II	PCR, qPCR, and microarrays followed by qPCR	DNA mutations allows checking DNA damage, which is essential for OSCC detection at its all stages

continue to be the primary approaches for analyzing these indicators. Microbiota propensity to illness and changes in behaviors, food, and host immunological response promote microbiota expansion. OSCC patients exhibit higher levels of *Tannerella forsythia*, *Porphyromonas gingivalis*, and *Candida albicans*. Mager et al. discovered rising amounts

TABLE 3.6 Clinical Significance of Oral Cancer Biomarkers

Clinical Significance	Oral Cancer Biomarker
Screening for oral squamous cell carcinoma	Salivary biomarkers such as L-phenylalanine, sphinganine, phytosphingosine, and S-carboxymethyl-L-cysteine
Differential diagnosis	Proteomic marker CLAC2
Recurrence potential marker	CD34 expression
Predicting prognosis and distant metastasis	Genomic biomarkers such as ITGA3 and ITGB4 expression
Predicts radioresistance in oral squamous cell carcinoma	Genomic markers such as VEGF, BCL-2, Claudin-4, YAP-1, and c-MET

TABLE 3.7   Potential Salivary Biomarkers for Oral Cancer Detection

α-Amylase	Quantitative proteomic analysis reveals decreased salivary amylase in oral cancer
Albumin	Serum and salivary levels of albumin as diagnostic tools oral malignant lesions
IL-1β	Interleukin-1 beta in unstimulated whole saliva is a potential biomarker of oral squamous cell carcinoma
IL-6 IL-8 TNF-α	Salivary IL-8, IL-6, and TNF-α are potential Diagnostic Biomarkers for Oral Cancer
Defensin-1	Defensin, a peptide can be detected in the saliva of patients with oral cancer
CD44 CD59	Salivary protein and solCD44 levels are a promising screening tool for head and neck squamous cell cancer recognition. Salivary indicators for the early detection of oral squamous cell cancer have been developed.
Resistin (RETN)	Saliva proteome profiling reveals potential salivary biomarkers for the detection of oral cavity squamous cell carcinoma
Haptoglobin (HAPβ) Hemopexin (HPX) Serotransferrin (TF) Transthyretin (TTR) Fibrinogenβ(FIBβ) α-1B glycoprotein (ABG)	Aberrant proteins in the saliva of patients with OSCC.
Annexin a2(ANXA2) Kininogen1(KNG1) Heat shock protein (HSPA5)	Saliva protein biomarkers to detect oral squamous cell carcinoma in a high-risk population
MMP-1 MMP-2 MMP-3 MMP-9 MMP-10 MMP-12 MMP-13	Tumor and salivary matrix metalloproteinase levels are strong diagnostic markers of oral squamous cell carcinoma. Serum and saliva collagenase-3 (MMP-13) in patients with oral lichen planus and OSCC
Mac-2 binding protein(M2BP) Myeloidrelatedprotein14(MRP14) Profilin Catalase Endothelin-1	Salivary proteomics for oral cancer biomarker discovery
Ga module complexed with human serum albumin (GA-HSA) Keratin-10 (K-10)	Detection of host-specific immunogenic proteins in the saliva of patients with oral squamous cell carcinoma

TABLE 3.7 *(Continued)*   Potential Salivary Biomarkers for Oral Cancer Detection

Keratin-2(K-2)	Identification of tumor-related proteins as
Galectin-7	potential screening markers by proteome
Cofilin	analysis—protein profiles of human saliva
CRP precursor	as a predictive and prognostic tool
Creatine kinase m-chain fatty-acid- binding protein (FABP)	
Myosin light chain 2, 3(MLC-2,3)	
Nucleoside diphosphate kinase (NDKA)	
Phosphoglycerate mutase 1 (PGAM1)	
Plakoglobulin (PG)	
Retinoic acid binding protein 2 (CRABP-2)	
Alpha-fetoprotein (AFP), carcinoembryonic antigen (CEA)	Expression and clinical significance of AFP and CEA in serum and saliva of patients with OSCC.
CA125 tissue	Saliva CA125 and TPS levels in patients
Polypeptide-specific antigen (TPA)	with oral squamous cell carcinoma
Basic fibroblast growth factor (FGF2)	The evaluation of basic fibroblast growth factor and fibroblastic growth factor receptor 1 levels in saliva and serum of patients with salivary gland tumor
Enolase-1	A screening test for oral cancer using saliva samples
Enzyme nicotinamide N-methyltransferase (NNMT)	basis for the development of a noninvasive diagnostic test for early-stage disease
Lactate dehydrogenase (LDH)	Increased level in Oral Cancer Patients

Other proteins: P53; Ki67; CA15-3; Cyclic D1; Cyfra 21.1; S100A2,7,9; STAT3; Statherin; Stratifin; Transferrin; Kallikerin-7; Glycolytic enzyme; IgG; Telomerase; Thioredoxin; Cystatin A; Truncated cystatin SA-I; Rostate specific antigen (PSA); Adenosine ;Transforming growth factor (TGF-1); Anti-oxidant like-1 (AOP-1); Deaminase (ADA); Serpin B3(SCCA1); 8-oxoquanine DNA glycosylase (OGG1)

of *Streptococcus mitis, Porphyromonas gingivalis,* and *Porphyromonas melanogenic,* suggesting that the particular microbiota may be used as a diagnostic signal in OSCC. Furthermore, the HPV and EBV genomic sequences have been found.

CONCLUSION

Research on oral cancer biomarkers in molecular biology and oncology focuses on identifying important biological molecules or markers that may be connected to cancer development, risk assessment, screening, recurrence prediction, indicating prognosis, indicating invasion/metastasis, and tracking therapeutic responses of cancer. It uses the enormous volumes of data produced by microarray technology to identify crucial biomarkers, which are then used in a variety of tools and databases to aid in the early diagnosis of various types of cancer.

REFERENCES

Arellano-Garcia ME, Hu S, Wang J, Henson B, Zhou H, Chia D, et al. Multiplexed immunobead-based assay for detection of oral cancer protein biomarkers in saliva. Oral Dis 2008;14:705.

Casciato DA & Territo MC (2012). Manual of clinical oncology (7th ed.). Wolters Kluwer Health/Lippincott Williams & Wilkins.

Casiato DA, Lowitz BB. Manual of Clinical Oncology (3rd ed.), 1983. Little, Brown and Co.

Chan DW, Sell S. Tumor Markers. In: Burtis CA, Ashwood ER, editors. Tietz's Textbook of Clinical Chemistry (2nd ed.), 1994. WB Saunders. 897–925.

Chang CL, Ho SC, Su YF, Juan YC, Huang CY, Chao AS, Hsu ZS, Chang CF, Fwu CW, Chang TC. DNA methylation marker for the triage of hrHPV positive women in cervical cancer screening: Real-world evidence in Taiwan. Gynecol Oncol. 2021;16:429–435.

Franzmann EJ, Donovan MJ. Effective early detection of oral cancer using a simple and inexpensive point of care device in oral rinses. Expert Rev Mol Diagn. 2018, 18, 837–844.

Lehto VP, & Pontén J (1989). Tumor markers in human biopsy material. Acta oncologica (Stockholm, Sweden), 28(5):743–762. https://doi.org/10.3109/02841868909092305

Manikantan NS, Balakrishnan D, Kumar AD, Shetty B. Tumor markers: At glance. Oral Maxillofac Pathol J. 2014;5:437–440.

Mishra A, Verma M. Cancer biomarkers: Are we ready for the prime time? Cancers (Basel). 2010;2:190–208. https://doi.org/10.3390/cancers2010190

Santosh AB, Jones T, Harvey J. A review on oral cancer biomarkers: Understanding the past and learning from the present. J. Cancer Res. Ther. 2016a;12:486–492. doi: 10.4103/0973-1482.176414

Santosh AB, Jones T, Harvey J. A review on oral cancer biomarkers: Understanding the past and learning from the present. J Can Res Ther. 2016b;12:486–492.

Schliephake H. Prognostic relevance of molecular markers of oral cancer – A review. Int J Oral Maxillofac Surg. 2003;32:233–45.

Shpitzer T, Hamzany Y, Bahar G, Feinmesser R, Savulescu D, Borovoi I, et al. Salivary analysis of oral cancer biomarkers. Br J Cancer. 2009;101:1194–1198.

Sun R, Juan YC, Su YF, Zhang WB, Yu Y, Yang HY, Yu GY, Peng X. Hypermethylated PAX1 and ZNF582 genes in the tissue sample are associated with aggressive progression of oral squamous cell carcinoma. J Oral Pathol Med. 2020;49:751–760.

Zhao Y, Hong XH, Li K, Li YQ, Li YQ, He SW, Zhang PP, Li JY, Li Q, Liang YL; et al. ZNF582 hypermethylation promotes metastasis of nasopharyngeal carcinoma by regulating the transcription of adhesion molecules Nectin-3 and NRXN3. Cancer Commun. 2020, 40, 721–737.

CHAPTER 4

Methodologies for Identifying Oral Cancer Biomarkers

Bhargav Shreevatsa[1], Viveka S[2], Sai Chakith M R[3],
Anisha S Jain[4], Chandan Dharmashekar[1] Ashwini Prasad[4],
Umamaheswari[4], Mahesh KP[2], Shiva Prasad Kollur[5],
Chandan Shivamallu[1]

[1]Department of Biotechnology and Bioinformatics, JSS Academy
of Higher Education and Research, Mysuru, Karnataka, India

[2]Department of Oral Medicine and Radiology, JSS
Dental College and Hospital, JSS Academy of Higher
Education & Research, Mysuru, Karnataka, India

[3]Department of Pharmacology, JSS Medical College, JSS Academy
of Higher Education and Research, Mysuru, Karnataka, India

[4]Department of Microbiology, JSS Academy of Higher
Education and Research, Mysuru, Karnataka, India

[5]School of Physical Sciences, Amrita Vishwa
Vidyapeetham, Mysuru, Karnataka, India

INTRODUCTION

Multiple approaches can be used to find the candidate biomarker. These
methods can lead to the discovery of biomarkers in the tumor cell, tumor
microenvironment, tumor-adjacent tissue, or pharmaceutical or therapeutic agent metabolism. To identify the candidate biomarker, a thorough
understanding of cancer biology is necessary (Sood et al. 2022).

DOI: 10.1201/9781032625713-4

Due to the practical and noninvasive mode of sample collection, salivary biomarkers have received more attention in the monitoring of the disease process and the therapeutic response as well as in the noninvasive identification of oral cancer. The number of studies reporting on unstimulated salivary constituents and speculating on their possible involvement in the field of oral cancer biomarkers has increased during the past two decades. Challenges in salivary biomarker research, however, have highlighted the necessity of standardizing saliva sample collection, enhancing sample processing and storage and reducing the large diversity in malignant and noncancerous individuals. Three salivary proteomic biomarkers and four salivary mRNA biomarkers have been found to be strongly linked with late-stage oral squamous cell carcinoma, according to research conducted by Brinkmann et al. (2011). Salivary proteomic biomarkers such as interleukin 1b (IL-1B), interleukin 8 (IL-8), M2BP, and mRNA indications like IL 8, S100P, SAT1, and IL 1B were detected at statistically significant levels. Biomarkers are discovered using a variety of molecular techniques, including DNA arrays, high throughput sequencing, polymerase chain reaction (PCR), gene expression arrays, restricted fragment length polymorphism, ribonucleoprotein immunoprecipitation, crosslinking immunoprecipitation, nuclear magnetic resonance, liquid chromatography, mass spectroscopy, enzyme assays, and immunohistochemistry. For accurate, valid, and trustworthy results on recently found biomarkers, studies aimed at identifying candidate biomarkers must statistically standardize their study design, sample population, and data processing (Chai et al. 2014).

The analysis and validation of the original hypothesis and findings, as well as the evaluation of additional information from the findings that can be used to guide clinical decision-making using analytic validity, clinical validity, and clinical utility, will be the main focuses of further testing after the discovery of candidate oral cancer biomarkers. Testing is done both before and after the creation of potential biomarkers. Pre-analytic validity refers to the management of the sample that will be examined using the new assay. The new assay's outcomes could be impacted by (1) the amount of time and storage circumstances between sample collection and processing; (2) the kind and length of fixation—or absence thereof; and (3) the amount of storage time and conditions after sample processing. Analytic validity is the evaluation of the technical features of the biomarker, which must satisfy particular standards and define the specificity and sensitivity of the test

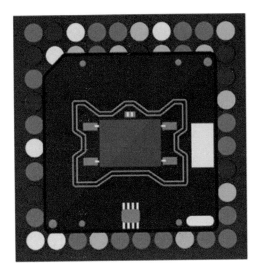

FIGURE 4.1   Microarray chip.

(Moore et al. 2011). The clinical validity of the biomarker will be examined after the development of the analytical validity of the assay. For a biomarker to be considered clinically viable, it must be able to accurately separate the total population of interest into two groups: Those who are more likely to suffer an event and those who are less likely to. A proposed biomarker's clinical value must be thoroughly investigated as the last phase in the development process (LOE). After that, the biomarker will be prepared for the use in direct patient care. The effectiveness and benefit-to-harm ratio of the biomarker are evaluated during this process. Despite the fact that thousands of biomarkers have been published in the literature, only a small number of tumor markers have been shown to be clinically useful.

Methods Employed in the Detection of Salivary Nucleic Acids
Microarray Technology
Researchers use microarray technology to look into the expression profiles of a large number of genes. A microarray is a collection of a number of miniature spots of particular DNA sequences (Oligonucleotides) located on a solid base (silicon chip, glass, and microscopic beads) (Ilhan et al. 2020). Each of these sequences is referred to as a probe, and it allows cRNA or cDNA from a sample to be hybridized. The working principle of microarray analysis is hybridization between two strands of nucleic acids through hydrogen bonds (Figure 4.1 and 4.2).

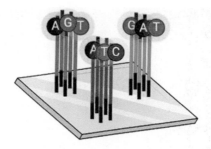

FIGURE 4.2   Oligonucleotides on a solid base.

First, the sequences in the research sample are combined with fluorescent dyes (cyanine 3, cyanine 5). The fluorescently labeled target sequences bind to the probe sequences and lead to a signal whose strength is measured (Figure 4.3). The number of photons released after stimulation with a laser of a specific wavelength is the unit of measurement. For each probe set, a digital image (Figure 4.4) is created, and then intensity values are obtained. Data analysis software like microarray suite software and plate readers are provided by microarray manufacturers to aid the creation of raw data. After that, background correction, spot filtration, quality control measures, normalization, and aggregation are applied, followed by gene expression identification and pattern recognition.

FIGURE 4.3   Hybridization between fluorescently labeled target sequences and probe sequences.

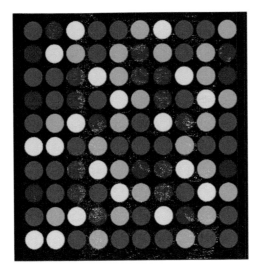

FIGURE 4.4   A digital image of microarray.

Quantitative Polymerase Chain Reaction

Quantitative PCR is a method for analyzing RNA that uses the quantity of mRNA in the sample to determine the amount of an expressed gene (underexpressed or overexpressed). It is necessary to reverse-transcribe the RNA sample into complementary DNA first (template). This procedure modifies the PCR using a heat cycler. Intercalated within the sample's fluorescent molecules, or "reporters," are sequence-specific probes. The fluorophore-labeled sequence combines with the complementary sequence to form a hybrid. The thermal cycler can illuminate each sample with a specific wavelength of light, and its sensors can detect the fluorescence emitted through the excited fluorophore. The instrument cycles through three stages of heating and cooling the samples. The nucleic acid double chain is isolated in the first stage; primer binding occurs in the second stage, and DNA polymerization occurs in the third stage. There are two main methods for quantifying gene expression: Relative quantification and absolute quantification.

Nanotechnology/Point-of-Care Platforms

Dr. Wong developed the oral fluid nano-sensor test, a relatively new point-of-care technology. This "lab on a chip" is a cost-effective, automated device that can detect around eight biomarkers in 15 mins. This system works based on salivary proteomes and transcriptomes being detected

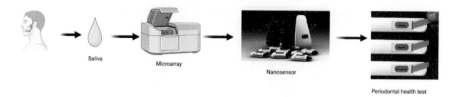

Saliva

Microarray

Nanosensor

Periodontal health test

FIGURE 4.5   Salivary diagnosis point-of-care technology.

electrochemically. Patients with an increased risk of oral cancer can be screened and then referred for biopsies (Figure 4.5).

Protocols in Biomarker Reporting and Evaluation

During the collection, processing, and storage stages, the consistency of biomarker biopsy specimens can be significantly influenced. Biomarker reporting protocols have been developed as a result of this to reduce the challenges in experimental outcomes and scientific results, as well as to ensure that all necessary information is included. Biospecimen reporting for improved study quality (BRISQ) (Moore et al. 2011), (REMARK) reporting recommendations for tumor markers (McShane et al. 2005), standards for reporting diagnostic accuracy (STARD) (Bossuyt et al. 2003), and minimum information about a microarray experiment are all standard protocols for biomarkers reporting (MIAME) (Taylor et al. 2007). BRISQ and REMARK are reporting criteria for pre-analytical and analytical issues related to potential prognostic factor studies in an organized and transparent manner.

STARD is a standard for publishing diagnostic tests, and MIAME is a set of reporting guidelines for microarray research. To determine the clinical utility of newly discovered biomarkers, a biomarker evaluation protocol has been developed. The Tumor Marker Guideline Committee of the American Society of Clinical Oncology proposed the tumor marker utility grading system (TMUGS) to aid in the critical evaluation of biomarkers. Evidence from a prospective clinical study to test the biomarker of interest or evidence from a meta-analysis or systematic review of well-conducted LOE II studies is required for the TMUGS protocol's highest LOE Level I (Hayes et al. 1996). The Level II studies use prospective clinical trials to provide evidence about a biomarker. Prospective clinical trial studies must be conducted at the highest LOE, namely LOE Level I, according to the revised critical evaluation protocol (Sargent et al. 2005).

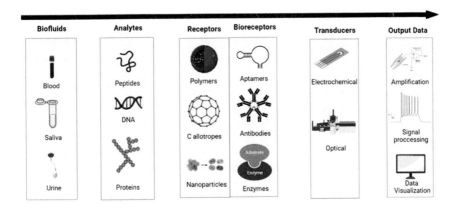

FIGURE 4.6   Components of a biosensor and its working principle.

Biosensors and Bioelectronic Platforms for the Detection of Oral— Cancer Biomarkers in Saliva

Commercially available ELISA kits have good sensitivity, but they have a long assay time by design. Biosensors have significant advantages in this regard, such as a faster time to respond and simpler protocols (Goldoni et al. 2021). A biosensor is a device that converts a biological event into a measurable signal. It is part of the larger category of chemical sensors. A biomarker (or a group of biomarkers) is chosen based on sensitivity and specificity criteria and then tested in the interested body fluid. The recognition element is the biomarker's binding site; it can be either natural bio-receptor or synthetic receptor, or a combination of both of them. The transducer then converts the biological signal generated by the binding event into a measurable signal via a detection mechanism (mass-based, optical, electrochemical, etc.) that is dependent on the biomarker (Jayanthi, Das, and Saxena 2017). Then the signal is amplified and subjected to several stages of signal conditioning and processing, either onboard the device or after it has been transmitted to the remote device (Jayanthi, Das, and Saxena 2017) (Figure 4.6).

CONCLUSION

Comprehensive clinical examinations, pricey biochemical analyses, and invasive biopsies continue to be the gold standard approaches for detecting oral malignancies. Early diagnosis may be possible through the discovery of biomarkers in biological fluids (blood, urine, and saliva). In this regard, the rapidly expanding fields of biosensing and point-of-care diagnostics may play a significant role in the early diagnosis of oral cancer. Numerous salivary indicators of oral cancer have been proposed, and saliva is becoming more and more popular as an alternative bio-fluid for noninvasive diagnostics.

REFERENCES

Bossuyt PM, Reitsma JB, Bruns DE, Gatsonis CA, Glasziou PP, Irwig LM, et al. The STARD statement for reporting studies of diagnostic accuracy: Explanation and elaboration. Ann Intern Med. 2003;138:W1–12.

Brinkmann O, Kastratovic DA, Dimitrijevic MV, Konstantinovic VS, Jelovac DB, Antic J, et al. Oral Squamous cell carcinoma detection by salivary biomarkers in a Serbian population. Oral Oncol. 2011;47:51–55.

Chai YD, Zhang L, Yang Y, Su T, Charugundla P, Ai J, et al. Discovery of potential serum protein biomarkers for lymph node metastasis in oral cancer. Head Neck. 2014;38:118–125.

Goldoni R, Farronato M, Connelly ST, Tartaglia GM, Yeo WH. Recent advances in graphene-based nanobiosensors for salivary biomarker detection. Biosens. Bioelectron. 2021a;171:112723.

Goldoni R, Scolaro A, Boccalari E, Dolci C, Scarano A, Inchingolo F, Ravazzani P, Muti P, Tartaglia G. Malignancies and biosensors: A focus on oral cancer detection through salivary biomarkers. Biosensors. 2021b;11:396.

Hayes DF, Bast RC, Desch CE, Fritsche H Jr., Kemeny NE, Jessup JM, et al. Tumor marker utility grading system: A framework to evaluate clinical utility of tumor markers. J Natl Cancer Inst. 1996;88:1456–1466.

Ilhan B, Lin K, Guneri P, Wilder-Smith P. Improving oral cancer outcomes with imaging and artificial intelligence. J. Dent. Res. 2020, 99, 241–248.

Jayanthi VSA, Das AB, Saxena U. Recent advances in biosensor development for the detection of cancer biomarkers. Biosens. Bioelectron. 2017; 91:15–23.

McShane LM, Altman DG, Sauerbrei W, Taube SE, Gion M, Clark GM; Statistics Subcommittee of the NCI-EORTC Working Group on Cancer Diagnostics. Reporting recommendations for tumor marker prognostic studies (REMARK). J Natl Cancer Inst. 2005;97:1180–4.

Moore HM, Kelly AB, Jewell SD, McShane LM, Clark DP, Greenspan R, et al. Biospecimen reporting for improved study quality (BRISQ). Cancer Cytopathol. 2011;119:92–101.

Sargent DJ, Conley BA, Allegra C, Collette L. Clinical trial designs for predictive marker validation in cancer treatment trials. J Clin Oncol. 2005;23:2020–2027.

Sood A, Mishra D, Kharbanda OP, Chauhan SS, Gupta SD, V Deo SS, Yadav R, Ralhan R, Kumawat R, Kaur H. Role of S100 A7 as a diagnostic biomarker in oral potentially malignant disorders and oral cancer. J Oral Maxillofac Pathol. 2022;26:166–172.

Taylor CF, Paton NW, Lilley KS, Binz PA, Julian RK Jr., Jones AR, et al. The minimum information about a proteomics experiment (MIAPE). Nat Biotechnol. 2007;25:887–893.

CHAPTER 5

Gene Expression Profile in OSCC

Viveka S[1], Bhargav Shreevatsa[2+], Kavana CP[2],
Chandan Dharmashekara[2], Ashwini Prasad[3],
Mahesh KP[1], Shiva Prasad Kollur[4],
Chandan Shivamallu[1]

[1]Department of Oral Medicine and Radiology,
JSS Dental College and Hospital, JSS Academy of Higher
Education & Research, Mysuru, Karnataka, India

[2]Department of Biotechnology and Bioinformatics, JSS Academy
of Higher Education and Research, Mysuru, Karnataka, India

[3]Department of Microbiology, JSS Academy of Higher
Education and Research, Mysuru, Karnataka, India

[4]School of Physical Sciences, Amrita Vishwa
Vidyapeetham, Mysuru, Karnataka, India

INTRODUCTION

In the management of OSCC, dental practitioners and pathologists confront two significant challenges: The heterogeneity of the disease and the absence of classical histological and clinical features. For instance, it will be difficult to determine which dysplastic disease, when a patient first presents to the clinic with one, will grow more or less violent over time based on the cytological markers of dysplasia.

For categorization of distinct phases of this illness, such as distinguishing between normal, dysplastic, and malignant cells, additional prognostic and predictive variables are required. In a recent article, this was

DOI: 10.1201/9781032625713-5

demonstrated. The purpose was to use DNA microarrays to look for differentially expressed genes that could best distinguish between normal, dysplastic, and cancer samples. To eliminate biological noise, all samples were acquired using laser capture microdissection. The investigation discovered several genes and expressed sequence tags from the normal and dysplasia analyses to be connected to cancer. DDB2, a damage-specific DNA protein, was discovered to be a possible biomarker for OSCC development (Braakhuis et al. 2003). This has the consequence that DDB2 may be a novel therapeutic aim for either treatment or prevention.

In OSCCs, chromosomal rearrangements and genomic instability have been detected using cytogenetic methods, fluorescence in situ hybridization, and comparative genomic hybridization. In Southeast Asian populations, including India, the clinical importance of the tumour suppressor genes and oncogenes in OSCC caused by cigarette smoking has been intensively investigated. Some of these studies emphasize the importance of conducting multicentric studies in different populations with well-defined tobacco habits to effectively translate biomarkers from the bench to the clinic, rather than simply translating a panel of biomarkers discovered in Caucasians to the Asian population without large-scale validations. Despite the fact that several genes (p16, p53, cyclin D1, p14ARF, Ets-1) have been linked to clinical and histopathological staging of OSCC, these markers provide little predictive or prognostic information for translation into a clinical setting. Given the vast heterogeneity of OSCCs, a single biomarker is unlikely to be useful for early detection of all OSCCs.

Oral cancer is caused by a multistep and multifocal tobacco-related process that includes field cancerization and intraepithelial, clonal spread. The most common oral premalignant lesion (OPL) is oral leukoplakia, which is frequently a precursor to oral squamous cell carcinoma (OSCC). Other OPLs and premalignant conditions with a well-defined risk of malignant transformation include erythroplakia, oral lichen planus, and oral submucous fibrosis. Prior to the development of invasive squamous cell carcinoma, OSCC progresses through a series of stages, from hyperplasia to varying degrees of dysplasia and carcinoma in situ. According to the "field cancerization" theory, an area of oral mucosa can be injured by repeated carcinogen exposure and then proliferate in a premalignant state. Individual subsites of the premalignant field develop as a result of additional genetic insults, leading to progression to frank carcinoma. According to this theory, even though relatively large areas of mucosa have a high proclivity for developing carcinomas, they may not show all of the traditional histological markers of malignancy (Datta, Patil, and Patel 2019).

The qualitative and quantitative patterns of groups of biomolecules (mRNA, proteins, peptides, or metabolites) in a cell, tissue, biological fluid, or organism are known as molecular signatures. The measurements should ideally be non-invasive and performed in a single readout to apply this concept to the discovery of oral cancer biomarkers. Braakhuis et al. propose the following model for head and neck carcinogenesis based on their genetic characterization of field cancerization. In the first phase, a stem cell in the oral mucosa's basal layer acquires a genetic alteration. When this altered stem cell divides into daughter cells, it forms a clonal unit of cells that all have the same DNA alteration and form a patch. As a result of the accumulation of more genetic changes, the patch transforms into an expanding field that pushes the normal epithelium aside. These fields may appear as oral patches, even though they are macroscopically invisible. Finally, cancer develops as a result of clonal selection within this field of preneoplastic cells.

The overexpression of several genes encoding matrix metalloproteinases (MMP1, MMP3, MMP7, MMP9, MMP10, MMP12, and MMP13), collagens (COL1A2, COL3A1, COL4A2, COL5A1, COL6A3, and COL11A1), C-X-C chemokines (CXCL1, CXCL2, CXCL6, CXCL9, CXCL10, and CX Repression of DNA binding inhibitors, protein phosphatase 1 regulatory (inhibitor) subunits, and numerous keratin genes—particularly those typically linked to OSCC and HNSCC, such as malignancies of the aerodigestive tract—accompanies these (KRT4, 13, 15, and 19).

Tobacco-Associated Proteomic Changes in Oral Cancer

Smoking is the primary method of tobacco use worldwide, and it is one of the top reasons for mouth cancer. Oral and oesophageal cancers have been linked to molecular changes caused by cigarette smoke and its constituents, according to studies. Few research has examined the molecular effects of different cigarette smoke and chewing tobacco, despite the fact that many epidemiological studies have connected smoking and chewing tobacco to oral malignancies. Because the constituents of the two types of tobacco differ significantly, the mode of action of these two tobacco products also differs significantly. Therefore, it is crucial to comprehend and distinguish between the molecular alterations caused by smoking and chewing tobacco, even when it comes to early diagnosis or screening of the population for threat assessment. Multiple groups have identified clinical biomarkers for OSCC using serum and saliva are examples of bodily fluids (Shaikh, Ansari, and Ayachit 2019). Adiponectin, cyclin D1, C-reactive protein, growth-differentiation factor (GDF 15), and cytokines such as interleukin 1 beta (IL-1), interleukin 6 (IL-6), and tumour necrosis factor-alpha have all been identified and are being studied for their potential diagnostic or prognostic use in OSCC. CYFRA 21–1 (cytokeratin fraction 21–1) has been reported as a prognostic marker for OSCC and other malignancies by Liu et al. There is currently a dearth of research describing possible biomarkers based on OSCC patients' tobacco use habits, despite the fact that multiple serum-based biomarker studies for the disease have been published.

Due to the composition differences, the pathobiology of OSCC caused by tobacco use, whether it be chewing or smoking, changes greatly. Plasminogen (PLG), fibrinogens (FGA, FGB, FGG), and actin (ACTB) are among the proteins whose abnormal expression contributes to malignant cell cycle progression in various cancers. While the majority of the proteins dysregulated in smokers are known to be involved in platelet degranulation, the majority of the proteins dysregulated in nonsmokers are unknown. These proteins are known to be dysregulated in lung cancer and have been linked to smoking-related epithelial diseases (Mohanty et al. 2021).

The complement pathway is an important part of humoral immunity, and the coagulation system is important for haemostasis, and their roles have been linked to a variety of diseases, including cancer. In the serum of OSCC patients who chewed tobacco, several hub proteins regulating this pathway were found, including fibrinogen (FGA, FGB, FGG), plasminogen (PLG), complement factors (C2, C5, C7, C9, C8A, C8G, CPB), coagulation factors (F2, F7, F13A1), and members of the serpins family (SERPIND1, SERPING1). Higher levels of plasma fibrinogen have been linked to a variety of cancers, including colon, pancreatic, nasopharyngeal, and oral cancer. Supplement factors and members of the serine protease family were discovered to be elevated in the OSCC chewer cohort. Studies have discovered higher amounts of serpin family G member 1 (SERPING1) and serpin family D member 1 (SERPIND1), both of which are known to play a key role in a complement cascade control, in the saliva of OSCC patients.

Immigration, invasion, and unchecked cell proliferation are all signs of OSCC development. Proteolytic enzymes like serine and cysteine proteases regulate such cellular processes, and serine proteases are regulated

TABLE 5.1   A List of Top Six Differentially Expressed Proteins Based on Tobacco Using Habits of OSCC Patients

Protein	Description	Functions
ITIH1	Inter-alpha-trypsin inhibitor heavy chain H1	Serine-type endopeptidase inhibitor action and calcium ion binding
SERPIND1	Serpin family D member 1	Serum glycoproteins encourage cancer cells' invasion and metastasis.
SERPINA4	Serpin family A member 4	Play a part in a variety of activities, including tumour development, angiogenesis, and inflammation.
SERPINA6	Serpin family A member 6	Protease inhibitor
ITIH2	Inter-alpha-trypsin inhibitor heavy chain H2	Heparin-binding activity and calcium ion binding
GSN	Gelsolin	Participate in cellular movement and act as a regulator throughout development

by members of the SERPIN (serine protease inhibitor) family. By using ELISA-based validation, the increase of SERPINA6 and SERPINF1 in the blood of OSCC patients who chewed tobacco was further validated. SERPINA6 is a secreted 52-kDa 1-glycoprotein that transports glucocorticoids and progestins through the bloodstream. Its elevated expression in OSCC patients' saliva has been well-documented. The 50 kDa glycoprotein SERPINF1 (also known as PEDF) has been shown to inhibit angiogenesis and metastasis and induce apoptosis (Bhat et al. 2021). It interacts with the serum albumin receptor and the F1 ATP synthase to prevent growth factor angiogenesis and reduced ATP synthesis, which causes endothelial cells to undergo apoptosis. These findings point to a significant role for SERPINF1 and SERPINA6 as biomarkers for OSCC patients who chew tobacco (Table 5.1).

CONCLUSION

The consumption of tobacco, whether smoked or chewed, has been linked to an increased risk of cancer, particularly oesophageal and oral malignancies. Regardless of cigarette use, the imbalance of immune signalling-regulated genes and muscle contraction-associated genes adds to the molecular changes for the tumorigenic phenotype. When oral cancer patients who smoked tobacco were compared to those who chewed tobacco, the bioinformatics analysis of dysregulated proteins found in our proteomic investigation demonstrated dysregulation of the antigen processing/presentation pathway and collagen synthesis.

REFERENCES

Bhat FA, Mohan SV, Patil S, Advani J, Bhat MY, Patel K, Mangalaparthi KK, Datta KK, Routray S, Mohanty N, Nair B, Mandakulutur SG, Pal A, Sidransky D, Ray JG, Gowda H, Chatterjee A. Proteomic alterations associated with oral cancer patients with tobacco using habits. OMICS: J Integr Biol. 2021;25(4):255–268.

Braakhuis BJ, Tabor MP, Kummer JA, Leemans CR, Brakenhoff RH. A genetic explanation of Slaughter's concept of field cancerization: Evidence and clinical implications. Cancer Res. 2003 Apr 15;63(8):1727–1730. PMID: 12702551.

Datta KK, Patil S, Patel K, et al. Chronic exposure to chewing tobacco induces metabolic reprogramming and cancer stem cell-like properties in esophageal epithelial cells. Cells. 2019;8:949.

Mohanty V et al.Molecular alterations in oral cancer between tobacco chewers and smokers using serum proteomic. Cancer Biomarkers. 2021;31:361–373.

Shaikh I, Ansari A, Ayachit G, et al. Differential gene expression analysis of HNSCC tumors deciphered tobacco dependent and independent molecular signatures. Oncotarget. 2019;10:6168–6183.

CHAPTER 6

Next Genome Sequencing in Oral Cancer

Bhargav Shreevatsa[1], Abhigna N[1], Viveka S[2],
Sai Chakith M R[3], Chandan Dharmashekara[1],
Siddesh V Siddalingegowda[4], Mahesh KP[2],
Shiva Prasad Kollur[5], Chandan Shivamallu[1]

[1]Department of Biotechnology and Bioinformatics, JSS Academy
of Higher Education and Research, Mysuru, Karnataka, India

[2]Department of Oral Medicine and Radiology,
JSS Dental College and Hospital, JSS Academy of Higher
Education & Research, Mysuru, Karnataka, India

[3]Department of Pharmacology, JSS Medical College, JSS Academy
of Higher Education and Research, Mysuru, Karnataka, India

[4]Department of Microbiology, JSS Academy of Higher
Education and Research, Mysuru, Karnataka, India

[5]School of Physical Sciences, Amrita Vishwa
Vidyapeetham, Mysuru, Karnataka, India

INTRODUCTION

Next-generation sequencing (NGS) is a massively parallel DNA sequencing method used for large-scale, ultra-high throughput and automated high-speed genome analysis. NGS has transformed the biological sciences by providing a less costly way of establishing nucleotides' order in whole

DOI: 10.1201/9781032625713-6

genomes or selected portions of DNA or RNA. NGS has several uses for studying biological systems at a higher level. The way that NGS handles tumor heterogeneity is another major element of the technology (Slatko, Gardner, and Ausubel 2018). Tumor heterogeneity, present in most solid tumors, makes detection and therapy incredibly difficult. Solid tumor oral squamous cell carcinoma (OSCC) is characterized by tumor heterogeneity, necessitating intensive genetic study. The NGS technique is crucial for identifying mutations with low variant allele frequency in this context (Kim, Lee, and Park 2020). The NGS technique may be used to sequence complete genomes or targeted portions of interest. OSCC analysis using NGS has also discovered several genomic changes and mutations with low variant allele frequency. Moreover, NGS is also becoming an important tool for identifying salivary gland malignances (Behjati and Tarpey 2013).

NGS is the most efficient approach for analyzing RNA and DNA in the area of functional genomics. This highly repeatable method may detect single nucleotide polymorphisms, linked gene variations, or spliced transcripts without requiring direct DNA or RNA input, as microarray approaches do. NGS may also be used on cDNA molecules for RNA sequencing by reverse transcription from candidate RNA and the building of sequencing libraries using enormous parallel deep sequencing NGS-based examination of possible biomarkers in saliva might also be utilized frequently as a simple and noninvasive test in OSCC patients. NGS has also been shown to be an effective approach for identifying new biomarkers in periodontal disease. Lately, NGS-based research has enabled the distinction of individuals with primary Sjögren's disease from those with nonsyndrome (Gasperskaja and Kučinskas 2017).

NGS also contributes significantly to discoveries, although its therapeutic applications have not yet reached their full potential. qRT-PCR or microarray has been used in several investigations to validate the presence of identified miRNAs. However, despite the launch of NGS more than a decade ago, only a few researchers have used it (Condrat et al. 2020).

Besides DNA sequencing, NGS may be used to analyze the RNA transcriptome (RNA-Seq). The transcriptome is the whole set of transcribed RNA in a sample, and sequencing allows for both relative gene expression analysis and the discovery of nucleotide polymorphisms. Until recently, gene expression microarrays, which rely on the hybridization and fluorescence of predesigned probes, were utilized in RNA expression research.

Gene expression microarrays, which have well-developed molecular technology and a vast body of research, have significant benefits over RNA-Seq. RNA-Seq may identify chromosomal translocations, fusion genes, differential splice variants, single nucleotide polymorphisms, and viral transcripts, among other transcriptome changes. RNA-Seq has been utilized in a variety of studies to explore gene expression and transcriptome variation in oral, esophageal, and oropharyngeal head and neck squamous cell carcinoma. RNA-Seq has been used to quantify differential gene expression, analyze gene ontologies to identify overrepresented, underrepresented, and dysregulated biochemical mechanisms, classify chromosomal translocations and subsequent fusion genes, assign differentially expressed novel mRNA splice variants, and search for HPV and other viral mRNA transcripts (Kukurba et,al 2015). Despite its great sensitivity and speed, real-time quantitative PCR (qRT-PCR) can only identify a small number of known sequences. NGS, in contrast to conventional sequencing techniques, enables RNA sequencing, targeted sequencing (which reveals important genes linked with cancer), and entire exome sequencing of malignancies. The way that NGS handles tumor heterogeneity is another major element of the technology. Tumor heterogeneity, which is present in the majority of solid tumors, makes detection and therapy incredibly difficult (Cseke and Larka 2016). The solid tumor OSCC is characterized by tumor heterogeneity, necessitating intensive genetic study. In this context, the NGS technique is crucial for identifying mutations with less variant allele frequency. NGS also contributes significantly to new discoveries, although its therapeutic applications have not yet reached their full potential. qRT-PCR or microarray has been used in several investigations to validate the presence of identified miRNAs. However, despite the launch of NGS more than a decade ago, only a few researchers have used it.

Future Technologies and Research Directions

The key to good clinical diagnostic tests is high sensitivity and specificity. Infections, periodontal, and pulpal diseases are just a few of the common inflammatory conditions that affect the oral cavity. The potential OSCC biomarkers in saliva may be affected by these nonneoplastic conditions. The majority of studies compared potential OSCC salivary biomarkers with healthy controls, and only a few studies confirmed the inflammatory conditions associated with OSCC. Such interactions may eventually lead to false positives as the level of inflammatory salivary biomarkers rises, overshadowing the potential of salivary biomarkers in OSCC detection. Salivary biomarkers, such as CA125, profiling, haptoglobin, transferring, and S100 calcium-binding protein, have been identified in other malignant lesions such as breast and lung cancers, and these biomarkers are also associated with OSCC detection. As a result, confirming the specificity of OSCC is critical for diagnosis. Furthermore, rather than a single biomarker, a panel of potential biomarkers can be used to make a precise diagnosis. As a result, extensive discovery and validation of novel biomarkers will be capable of revolutionizing the field of oral tissue cancerous and noncancerous performance (Zhang et., al 2021).

Oral cancer candidate biomarkers should be identified and classified in the fields of screening, differential diagnosis, recurrence prediction, prognosis, therapeutics, and metastasis in future studies. These potential biomarkers will be useful in predicting clinical outcomes and formulating dental public health measures. Tumor markers cannot be utilized as a solitary diagnostic tool; however, they can be used in conjunction with hematoxylin and eosin staining in regular histology. They can also be used in conjunction with diagnostic procedures to confirm and grade the malignancy. Furthermore, by integrating many tumor markers, we can improve specificity and sensitivity in the follow-up of one kind of malignancy, the underlying process that leads to an aggressive phenotype, which is yet unknown (Nguyen et al. 2020).

As a result, biomarkers that can predict prognosis are desperately needed. The creation of biomarkers aimed at assessing the success of oral cancer medication therapy would be extremely helpful in determining therapeutic efficacy. Researchers are encouraged to apply novel research methods that comply with the concepts of analytic validity, clinical validity, and clinical usefulness, as well as biomarker reporting and assessment standards. To better comprehend the diverse cell population of cancer tissue and the host immune response to the cancer cell population, more research to create biomarkers is required. It is suggested that research be focused on distinguishing between biomarkers for cancer diagnosis and treatment targets. Early detection of oral cancer can assist to minimize patient morbidity and death, hence developing clinically viable candidate biomarkers with higher clinical usefulness values for oral cancer screening is highly suggested (Almangush, Leivo, and Mäkitie 2021). The areas that are critical in the future of oral genomic research include new approaches to identify different features in DNA sequence, the analysis of gene and protein expression and the determination of the relationship between genotype and phenotype, and the identification of the patterns in the genetic variation in populations

The field of tumor markers has a bright future ahead of it. Human diseases will be classified based on molecular rather than morphological analysis as genomic and proteomic technology advances. This will be accomplished using techniques such as laser capture microdissection for tissue and cell procurement, as well as combining genomic and proteomic analysis. Using unique gene or protein profiles containing multiple biomarkers, the disease can be diagnosed early. For parallel testing, traditional ELISA or antibody-based protein chips can be used to analyze panels of protein biomarkers. Furthermore, as a result of genomic and proteomic discoveries, many more diagnostic tests will be developed. Biochip development will grow much faster in the future than the rest of the diagnostic industry, which will include DNA, RNA, and protein chips. All types of samples, including tissues, cells, and bodily fluids, will be examined. Integrated diagnostic tools will be used, which combine these methods with molecular imaging techniques. Finally, bioinformatics will link scientific data to clinical information to provide better, more comprehensive patient care. New discoveries will be rapidly translated from the laboratory to the patient's bedside. Laboratory testing and, as a result, laboratory diagnosis are becoming increasingly important in the integrated healthcare delivery system, thanks to advances in proteomic technology (Huang 2001).

Rapid advancements in flexible electronics and miniaturized technologies have enabled the creation of modular and scalable bioelectronic systems for early cancer screening using easily available bodily fluids. Saliva's value as a noninvasive source of relevant biomarkers reflecting illness load and progression has been demonstrated by the discovery of alterations in saliva that indicate disease progression. Saliva has a number of disadvantages, including quick biofouling on biosensor surfaces, the influence of interferents contained in saliva at various quantities, and the existence of a highly energetic environment orally. Nonetheless, researchers have been able to overcome most of the current limitations in salivary diagnostics because of the development of powerful bioelectronic techniques.

The relevance of saliva as an easily available and constantly renewed biofluid has been emphasized to spur further research on salivary biomarkers, whose early detection might enhance patients' quality of life and survival rate. However, before these compounds may be classed as potential biomarkers, their origin must be confirmed. New translational and transitional studies are required as the initial stage in marker evolution research to transfer this fundamental knowledge into biomarkers to prevent disease and therapeutic usage. We won't be able to reduce potential systemic and random mistakes in clinical research until we do specialized investigations on the biological and technological variability of salivary biomarkers. Clinical investigations on the diagnostic, predictive, and prognostic performance of saliva are needed. The final step before saliva biomarkers can be used in clinical practice is completing the well-designed clinical trials on their diagnostic, predictive, and prognostic performance. The use of sophisticated biosensors targeting biofluids is now limited to research settings due to the lack of disease-specific markers that can ensure absolute sensitivity and specificity. As a result, the biomarkers mentioned in the study should be regarded as potential biomarkers that should be targeted in future research rather than biomarkers that have matured enough to be used in clinical practise (Aro et al. 2017). In the previous decade, a large number of salivary biomarkers have been presented for cancer screening or early detection, indicating a high interest from the medical community. However, to draw substantial connections between the results, it is critical to confirm the findings acquired in pilot research among broader demographic groupings.

Given the probability of frequent false positive and false negative events, doctors are unlikely to deploy new biosensing techniques based on salivary biomarkers in clinical practice without extensive validation. Furthermore, because saliva is a biofluid with a highly energetic composition, confounding variables should be considered carefully while doing salivary analysis, collection and sampling techniques should be as standardized and automated as feasible to minimize the presence of confounders. Wearable chemical biosensors might be useful in this case for tracing the concentration of certain biomarkers over a lengthy period of time, while advanced statistical algorithms could be utilized in the postprocessing phases to mitigate the problem. Wearable intra-oral bioelectronic platforms and portable POC devices are projected to gain clinical acceptance in the near future, since they provide significant benefits over standard procedures and laboratory equipment. Laboratory procedures need specialized equipment as well as qualified individuals to operate these complicated systems. This quickly translates into increased expenditures for each analysis and a longer time-to-response, making these procedures complex and inefficient. Advanced biosensing devices may now be manufactured at a fraction of the cost of costly laboratory equipment because of rapid advances in electronics and nanotechnologies. Because salivary biomarker concentrations are frequently lower than those found in other biofluids, typically in the ppb-ppt range, ultrasensitive biosensors are required to successfully detect them Cheng et., al 2014.

Limit of detection and sensitivity matches, for the most part, eliminate the limitations of previous approaches while giving a simpler operation that can be performed by the user with minimum effort. Furthermore, novel materials are being investigated that potentially provide "green" alternatives to standard materials throughout the production process, allowing for better long-term diagnostic tools. To transform research efforts into substantial practical contributions, scalable, low-cost, large-scale production technologies are in high demand. Integration of miniature wireless communication units into these integrated bioelectronic devices is also a significant enabler for smart linked systems in the future. In the age of customised medicine and cloud-based storage of sensitive data, it is critical for patients to retain ownership of their own data through improved data collecting and security technologies. Overall, the ability to exchange clinical data with doctors might lead to more accurate forecasts using breakthrough big data techniques (Schmidt, Marques, and Botti 2019). As a result, improved diagnosis, prognosis, and therapy would be possible. Intraoral biosensors are expected to play a key role in continuous real-time data collecting for the identification of potentially malignant indicators in oral cancer. Because saliva comes into close touch with premalignant or malignant cells, oral cancer is the ideal candidate for detection with wearable sensors inserted in the mouth cavity.

CLINICAL APPLICATION OF NEXT-GENERATION SEQUENCE

The factors influencing the spread of lung, ovarian, prostate, and bile duct cancer have been identified through NGS-based research. A routine, easy-to-use, noninvasive screening for patients with OSCC might also be performed using an NGS-based examination of potential biomarkers in saliva.

Additionally, GS has been effective in identifying new biomarkers for periodontal disease. The benefit is that the majority of the NGS platforms now on the market use the same parallel sequencing method for clonally amplified DNA molecules. Several studies have shown that the use of NGS for OSCC sequencing can produce accurate findings at high throughput. Researchers may be able to assess genetic links between various tumor clones more precisely and consistently with the use of the combined analysis of these data (Li, Liu, and Du 2022).

CONCLUSION

NGS is being developed as a significant research tool for evaluating genomic changes in a range of human disorders. The benefit is that the majority of the NGS platforms now on the market use the same parallel sequencing method for clonally amplified DNA molecules. NGS can play a bigger part in clinical practice as our understanding of its utility advances, but only if some of its drawbacks are resolved. Education about its uses, the availability of precise bioinformatics tools to evaluate the massive amounts of generated data, and improving the technical skills and knowledge of laboratory operators are urgently needed. Overall, NGS is a key finding that will aid in the accurate diagnosis of diseases and the prompt application of effective treatments.

REFERENCES

Almangush A, Leivo I, Mäkitie AA. Biomarkers for immunotherapy of oral squamous cell carcinoma: Current status and challenges. Front Oncol. 2021;11:616629. doi: 10.3389/fonc.2021.616629

Aro K, Wei F, Wong DT, Tu M. Saliva liquid biopsy for point-of-care applications. Front Public Health. 2017;5:77. doi: 10.3389/fpubh.2017.00077

Behjati S, Tarpey PS. What is next generation sequencing? Arch Dis Child Educ Pract Ed. 2013;98(6):236–238. doi: 10.1136/archdischild-2013-304340.

Cheng YSL, Rees T, Wright J. A review of research on salivary biomarkers for oral cancer detection. Clin Trans Med. 2014;3:e3. https://doi.org/10.1186/2001-1326-3-3

Condrat CE, Thompson DC, Barbu MG, Bugnar OL, Boboc A, Cretoiu D, Suciu N, Cretoiu SM, Voinea SC. miRNAs as biomarkers in disease: Latest findings regarding their role in diagnosis and prognosis. Cells. 2020;9(2):276. doi: 10.3390/cells9020276.

Gasperskaja E, Kučinskas V. The most common technologies and tools for functional genome analysis. Acta Med Litu. 2017;24(1):1–11. doi: 10.6001/actamedica.v24i1.3457.

Handbook of Molecular and Cellular Methods in Biology and Medicine Edited By Leland J. Cseke, Ara Kirakosyan, Peter B. Kaufman, Margaret V. Westfall. ISBN 9780429112386–735 Pages Published April 19, 2016 by CRC Press

Huang R-P. Detection of multiple proteins in an antibody-based protein microarray system. J Immunol Methods. 2001;255:1–13. doi: 10.1016/S0022-1759(01)00394-5.

Kim S, Lee JW, Park YS. The application of next-generation sequencing to define factors related to oral cancer and discover novel biomarkers. Life (Basel). 2020;10(10):228. doi: 10.3390/life10100228.

Kukurba KR, Montgomery SB. RNA sequencing and analysis. Cold Spring Harb Protoc. 2015;2015(11):951–969. doi: 10.1101/pdb.top084970.

Li P, Liu S, Du L et al. Liquid biopsies based on DNA methylation as biomarkers for the detection and prognosis of lung cancer. Clin Epigenet. 2022;14:118. https://doi.org/10.1186/s13148-022-01337-0

Nguyen TTH, Sodnom-Ish B, Choi SW, Jung HI, Cho J, Hwang I, Kim SM. Salivary biomarkers in oral squamous cell carcinoma. J Korean Assoc Oral Maxillofac Surg. 2020;46(5):301–312. doi: 10.5125/jkaoms.2020.46.5.301.

Schmidt J, Marques MRG, Botti S et al. Recent advances and applications of machine learning in solid-state materials science. Npj Comput Mater. 2019;5:83. https://doi.org/10.1038/s41524-019-0221-0

Slatko BE, Gardner AF, Ausubel FM. Overview of next-generation sequencing technologies. Curr Protoc Mol Biol. 2018;122(1):e59. doi: 10.1002/cpmb.59

Zhang M, Chen X, Chen H, Zhou M, Liu Y, Hou Y, Nie M, & Liu X (2021). Identification and validation of potential novel biomarkers for oral squamous cell carcinoma. Bioengineered, 12(1):8845–8862. https://doi.org/10.1080/21655979.2021.1987089.

Index

For Product Safety Concerns and Information please contact our
EU representative GPSR@taylorandfrancis.com Taylor & Francis
Verlag GmbH, Kaufingerstraße 24, 80331 München, Germany